ISBN 978-1-331-97117-7
PIBN 10262217

1 MONTH OF
FREE
READING

at

www.ForgottenBooks.com

By purchasing this book you are eligible for one month membership to ForgottenBooks.com, giving you unlimited access to our entire collection of over 700,000 titles via our web site and mobile apps.

To claim your free month visit:

www.forgottenbooks.com/free262217

CATALOGUE

OF THE

CASES OF BIRDS

IN THE

DYKE ROAD MUSEUM,

BRIGHTON.

*Giving a few Descriptive Notes, and the Localities in which the
Specimens were obtained.*

BY

E. T. BOOTH.

Brighton :

GEORGE BEAL, PRINTER, BOOKSELLER AND STATIONER,

207, WESTERN ROAD.

—

1876,

INTRODUCTION.

As a Catalogue does not need a Preface, I will simply state, by way of introduction, that all scientific arrangement has been given up as hopeless in a collection where the chief object has been to endeavour to represent the birds in situations somewhat similar to those in which they were obtained ; many of the cases, indeed, being copied from sketches taken on the actual spots where the birds themselves were shot.

The few notes and facts I have recorded are solely the result of personal observation, and with two or three exceptions (all noted), not a book of reference has been opened.

Those who expect to find a long list of rarities will, I am afraid, be sadly disappointed, as, in order to avoid exhibiting or describing a specimen with which I was only acquainted by hearsay, I have restricted the collection entirely to birds that have fallen to my own gun during various excursions in the British Islands.

CATALOGUE OF BIRDS.

GOLDEN EAGLE.—(IMMATURE.)

Case 1.

The specimens in this case (in conjunction with those in 306) show some of the various stages of plumage exhibited by this Eagle during its progress towards maturity.

It is probable that birds of this species are five or six years old before they assume the full mature plumage. Golden Eagles vary considerably, but they should not, I imagine, be considered perfectly adult till all signs of white have disappeared from the tail, and also till the feathers on the legs have become a warm dark brown or rust colour.

The two specimens on the right, a male and female, I should judge to be two years old, and the remaining bird, a male, in the last stage before assuming the perfect adult plumage.

The birds were all trapped in the Northern Highlands in the Spring of 1878.

WHITE-TAILED EAGLE.

Case 2.

Though banished from numbers of eyries where it was formerly in the habit of breeding, the White-tailed Eagle still holds its own on the Western Coast of Scotland.

The male and female are here shown with their nest and eggs.

The case is copied from a sketch made in the Hebrides. The female was shot and the male trapped within a few miles of the same spot in the Spring of 1877.

SWIFT.

Case 3.

This is the last of the Swallow tribe to visit us in the spring and the first to depart in the autumn.

Swifts are generally supposed to nest in holes under the eaves of houses or churches, but where suitable places of this description are wanting, they do not hesitate to make use of fissures and cracks in the face of cliffs or precipices.

The specimens in the case were obtained at the Cromarty Rocks, on the north-east coast of Scotland, in June, 1869.

This bird and the House Martin are here found in great numbers during the summer.

GOATSUCKER.

Case 4.

The Goatsucker, or Fern Owl, though unknown to many from its nocturnal habits, is a common bird from north to south.

I have noticed it as particularly numerous in Sussex, Norfolk, and Ross-shire.

It arrives in this country in May, and usually takes its departure as soon as the rough weather in the autumn commences.

I have often at dusk seen several flying about the streets and round the chimneys, in towns on the southern and eastern coasts, during a gale of wind, previous to their departure.

The nest is placed in an open spot on a heath or moor, or on the middle of a foot-track through a wood. The birds rest quietly on the ground by day, and as soon as dusk sets in commence their jarring note, from which they derive the name of Nightjar. Their food consists mainly of moths and night-flying insects. They have a most capacious mouth, and are provided with a serrated claw, which is supposed by country people to be intended for combing the scales or down of the moths from their whiskers.

The birds were obtained on the Hill of Tarlogie, near Tain, in Ross-shire, in June, 1869.

SWALLOW.

Case 5.

There are several weeks in the spring during which Swallows may be daily observed landing on our shores.

The first arrivals usually make their appearance early in April; and as late as the 20th of May, when out in the Channel, I have met with hundreds still crossing.

Should boisterous weather set in shortly after their arrival, they suffer greatly from the effects of the cold and wind. The weather in the second week in June, 1871, in the east of Norfolk, was unusually severe for that time of year, and the unfortunate Swallows and Martins were seen in hundreds sheltering from the storm under the hedges and banks. I brought in several quite benumbed by cold, and after placing them in a warm room for some hours, they were enabled to fly off in search of food.

Swallows and Martins may frequently be observed, when flying over a river or pond, dipping into the water to drink. This operation is generally easily effected. I, however, noticed several hundreds of these birds lose their lives in attempting the same thing a few years back in the east of Sussex.

The water had been drawn off from a large fish-pond, and although the surface still retained its usual appearance, it was in reality nothing more than a thick black mud of about the consistency of treacle.

The old saying, that "one fool makes many," certainly referred to the poor Swallows, for no sooner had the wing of one unlucky bird been caught by the mud while skimming too closely over the surface, than the struggles of the sufferer brought scores to the spot, and within an hour or two the mud was dotted all over with hundreds of dead and dying victims. Some of those nearest the shore were reached with landing-nets, and after being cleansed from the mud in fresh water, and placed in the sun for a short time, were enabled to dry their feathers and make good their escape.

A few Swallows frequently remain long after the main body have left our shores for a warmer climate. These, I believe, are for the most part young birds, either too backward or weak to attempt the journey. They are occasionally noticed as late as Christmas, if the weather continues open, but as they are generally lost sight of after a few days' frost, it may be supposed that they have at last succumbed to cold and hunger.

The specimens in the case were obtained in Sussex, during the summer of 1870.

GOLDEN PLOVER.—(SUMMER.)

Case 6.

Golden Plovers, with black breasts, usually arrive at their breeding-quarters in the Highlands at the end of April or beginning of May; the time, however, varies with the state of the weather. In the spring of 1867 they made their appearance rather early in the north-western part of Perthshire, and had taken up their summer quarters, when a heavy fall of snow again drove them south, and only about half a dozen pair returned and nested on ground where hundreds are generally found.

It is a wonder how the eggs and young, in such exposed spots, are enabled to withstand the effects of the frost and snow. I have frequently observed the newly-hatched young on the hills in close proximity to snow-drifts of twenty or thirty feet deep.

The bird is usually obtained in its finest plumage

immediately it arrives at its breeding-quarters ; soon after commencing nesting, white feathers begin to show among the black, and its handsome appearance is consequently spoiled.

Like the Green Plover or Peewit, several pairs are commonly found nesting in company. When anyone approaches the neighbourhood of their nests they show the greatest concern, never ceasing calling and flying round till the cause of their annoyance has disappeared.

The specimens in the case were obtained in Glenlyon, in Perthshire, in the beginning of June, 1867.

GOLDEN PLOVER.—(Autumn.)

Case 7.

This is the plumage in which the bird is best known to those living south of the Tweed

By the time the young are strong enough to leave the hills, the old birds have mostly assumed their autumn dress, and joining together in small flocks, they make their appearance on the shores of the Scotch firths. I have, however, occasionally met with specimens, as late as the middle of September, which appeared, when on wing, to be almost in full summer plumage ; but if closely examined, it would be found that the black feathers in their breasts were thickly interspersed with white.

Some few young birds, singly or in small parties, occasionally wander as far south as Norfolk, or even Sussex, by September ; but being remarkably tame, they generally fall victims to the first gunner they approach, being easily enticed within shot by an imitation of their own note, however badly executed.

During severe weather they may be found congregated* to the number of several thousands, generally frequenting tidal mudbanks, and retiring at high water to the adjoining marshes. It is at this season, while feeding on the mud, that they occasionally offer chances of which the punt-gunners in the neighbourhood are not slow to avail themselves—as many as fifty, sixty, and seventy, being frequently obtained at a shot.

The specimens in the case were shot on Breydon mudflats, in September, 1871.

SAND MARTIN —(IMMATURE.)
Case 8.

During the early part of the autumn, large numbers of Sand Martins (mostly young) may be observed, particularly during wet weather, settled on the banks and among the reeds that surround the broads and large pieces of water in the eastern part of the island. The case is intended to represent the birds in such a position.

The specimens were obtained near Shoreham, in Sussex, in September, 1875.

---

* Those who have never examined any old works on fowling or gunning, may possibly not be aware that the sportsmen of former days had special terms for the flocks of every description of wildfowl, in the same manner as we speak of a covey of partridges, a bevy of quail, or a wisp of snipe. It used to be
A congregation of Plover,
A herd of Swan, &c. &c.
Folkard on Wildfowling gives a full account on pages 5 and 6.

WREN.

Case 9.

It is needless to say much about this familiar little bird. *Jenny* Wren is almost as well known as *Cock* Robin.

It may possibly, however, have escaped the notice of some observers, that these little birds have a singular habit of roosting together in great numbers during cold weather.

I have repeatedly counted as many as ten or a dozen, just at dusk. flying one after another into a hole in a haystack, or in the thatch of some outbuilding.

The specimens in the case were obtained at Portslade, near Brighton, in June, 1874.

SAND MARTIN.—(Mature.)

Case 10.

We frequently have severe weather in the spring after the arrival of this poor little traveller.

The Sand Martin, however, appears to be a remarkably hardy bird, as I have sometimes noticed thousands huddled together on the reeds in the broads of our eastern counties during a snow-storm early in the spring, and apparently none the worse, should the sun break through on the following day.

When I lived in Glenlyon, in Perthshire I was surprised to notice one season that no Sand Martins nested in the banks of an island in the Lyon, where I had observed them the previous year. We were, however, visited by a very high flood in June, which completely covered the whole of the island, and caused considerable damage in the district by sweeping away both cattle and sheep.

I well remember landing, with a cast of the phantom minnow, the carcase of a fine ram, which was coming down the river with the first of the spate as I was returning from fishing.

The specimens in the case were obtained in Norfolk, in May, 1870.

HOUSE MARTIN.—(MATURE.)

Case 11.

To by far the greater number of the British public, Swallows, Swifts, and Martins are generally known by the name of Swallows.

Those, however, who take an interest in our small visitors may easily distinguish the little bird with blue-black plumage and broad white bar across the tail, and remember that scientific naturalists have bestowed on that tiny traveller the euphonious title of "Hirundo Urbica," while to humbler observers like ourselves it is simply known as the House Martin.

Most people welcome the arrival of these familiar visitors, and afford them protection when nesting under the eaves of their houses. They will, however, when taking a fancy to the corner of a window-frame, occasionally become a most persevering nuisance, insisting to fix their nursery, with all its accompanying dirt, to the glass of the window, even after receiving several forcible hints that the situation is unsuitable.

Like all the rest of the family, they are not only

perfectly harmless to gardeners and farmers, but they confer an inestimable boon on all, by ridding the air of millions of nocuous insects.

If any of our British birds require a law to protect them, there are none, in my opinion, more worthy of it than the Swallow tribe.

It was lately stated in print that a certain firm of plumassiers had given out an order for a hundred dozen Swallows and Martins.   Such wanton destruction of a useful bird ought certainly, if possible, to be put a stop to.

House Martins, although generally nesting, as their name implies, round the dwellings of man, may be occasionally found building amongst rocks and cliffs.

The specimens in the case were taken, together with their nest, at the Cromarty Rocks, in July, 1869.

DARTFORD WARBLER.

Case 12.

This is by no means an uncommon, though a decidedly local species.   I have, however, only met with it in the south-eastern counties.

During the summer, Dartford Warblers may generally be found in most of the large patches of furze that are scattered over the South Downs, though, being remarkably shy, they are liable to escape observation, as, on the slightest signs of danger, they immediately seek the shelter of the bushes.

In the winter they seem to be of a roving disposition, as I have met with them several times among the stunted thorn-bushes and straggling furze on the beach between Eastbourne and Pevensey, and when rabbit-shooting further inland, I have noticed them occasionally driven out by the beagles from cover, where no one would ever imagine they would be found.

The nest is small, and very artfully concealed. If deprived of their first nest, one pair will continue attempting to rear a brood till late in the season, even after being robbed of three or four sets of eggs.

They feed their young generally on the body of a largish yellow moth. I observed several pairs carrying a white substance in their mouths to their nests, which I could not make out, and on shooting one bird from each of two nests I discovered that the food was identical in both cases. The wings of the moth were removed, so I was not entomologist enough to name the species, but I observed that the birds hunted for their prey among the lower part of the stems of the furze.

The specimens in the case were taken near Brighton, in July, 1869.

SEDGE WARBLER.

Case 13.

This lively little bird is found from north to south, wherever there are localities adapted to its habits.

It seems as noisy and as much at home in the reeds round a Highland loch as it does when met with in the fens of Cambridge or the broads of Norfolk.

The nest is generally placed at no great distance from water, either among the roots of the sedges, on a rough bank, or against the stump of a tree.

The specimens in the case, both old and young, were obtained near Heigham Sounds, in Norfolk, in July, 1871.

GRASSHOPPER WARBLER.

Case 14.

Though frequently found in the neighbourhood of water, this Warbler is by no means so aquatic in its habits as the more common Reed and Sedge Warblers. Several pairs breed round most of the broads in the east of Norfolk, and I have also discovered their nests in hayfields and bramble-covered banks in the more southern counties.

When shooting in the Nook, at Rye, in Sussex, early one morning in May, 1858, I found the samphire and other small weeds that grow on the mudbanks completely swarming with Grasshopper Warblers. They had evidently only just landed, and were on the point of making their way inland. There must have been several hundreds in a small patch of weed of a dozen or twenty acres. There were probably some other small birds of passage among them, but two shots which I fired into the weeds produced about half a dozen, all of which were of this species.

I am unable to account for so many being found together, as I have noticed that our spring migrants arrive, for the most part, singly or in small detached parties, large numbers seldom being observed flying in company.

Though very difficult to catch a glimpse of during the day, even near their nesting-quarters, they may generally be seen about daybreak singing on some high reed or branch of a tree. The slightest sign of danger, however, is sufficient to cause them to drop like a stone into the thick cover, where they quietly remain, creeping about like a mouse till the place is again quiet.

One of the specimens in the case was shot near Brighton, in May, 1868; the other in the marshes near Hickling Broad, in Norfolk, in May, 1873.

TURNSTONE.—(SUMMER.)

Case 15.

Turnstones in their summer dress may usually be looked for along the coast about the middle of May.

Some few of these birds occasionally remain the whole of the summer in Great Britain. This is particularly the case along the rocky parts of the coast, commencing with the Fern Islands, and terminating opposite the Bass Rock, in the Firth of Forth.

Their being found at this time of the year has led some writers to state that they breed on our shores; but up to the present time, I believe, no authenticated eggs have been taken in this country.

I have on two or three occasions shot a few of these late-staying birds on purpose to examine them, and have always found that, although in what might be styled summer plumage, they were never nearly so perfect in colour as the birds that pass along the coast in May, giving one the impression of their being either backward or sickly.

They feed on the small worms and salt-water insects which are found on the mudbanks and rocks they frequent.

It is a common habit with these birds to turn over seaweed, stones, and dead fish, or other refuse under which their food might be concealed.

Two of the specimens in the case were obtained on

Gullane Links, in East Lothian, in May, 1867, and the remainder on Breydon mudflats, in Norfolk, in May, 1871.

REED WARBLER.

Case 16.

Scientific naturalists declare we have in this country two distinct species of Reed Warblers, but whether this is the case or not, I leave to wiser heads than mine to decide.

Wherever reeds are abundant this bird is sure to be met with, either along the banks of rivers and ponds, or in the large beds that are found in the neighbourhood of the broads of Norfolk and Suffolk and adjoining counties.

The nest is usually attached to three or four stems of the reed; and if rocking is a luxury to the young birds, they must, certainly during rough weather, have a particularly happy time of it, as their cradle sways backwards and forwards with every breath of wind.

Like their neighbour the Sedge Warbler, they are remarkably noisy, though not extremely melodious songsters. During the day both species confine themselves to an occasional cackling note, evidently reserving their harmony for the evening concert, which usually commences as soon as the sun gets low. Hickling Broad, in the east of Norfolk, is one of the spots where this may be heard to perfection any fine evening in June.

The din that is caused by several hundreds of these birds singing and chattering at the same time, together with the croaking of the frogs, the jarring of the Night

Hawks, and the drumming of the Snipes, is perfectly deafening, and would never be credited by those who have not heard it. By about **11 p.m.** the greater part of the performers are quiet, but the slightest sound, even the slushing of a large pike on the look-out for his supper, is enough to make them break out again in full chorus.

During cold and stormy weather they remain remarkably silent, hardly a sound, except the occasional scream of a *C*oot or Moorhen, being heard through the swamps, to break the monotony of the sighing of the wind through the reed-beds and the splash of the rain in the open water.

The old birds, with their young, were obtained on Heigham Sounds, in Norfolk, in June, 1871.

MEADOW PIPIT.

Case **17.**

This is one of the commonest of our British birds. Although several of these Titlarks remain with us through the winter, their numbers are considerably augmented by fresh arrivals in the spring. Any still foggy morning, from the middle of March till well on in April, they may be noticed landing on the south coast, singly and in small parties, from daybreak till nine or ten o'clock. For a day or two they may be observed in numbers about the banks of streams and salt-water pools near the sea beach; but with a change of weather, they soon proceed inland, and scatter themselves over the country.

About October there seems to be a general movement of these birds along the south coast, their line of flight being from east to west; but whether they are about to cross the Channel, or what the object of their flight may be, I am unable to say.

The persecution that this unfortunate Pipit undergoes from the various smaller Hawks in the Highlands ought to tend to keep down their numbers. Merlins, Sparrow Hawks, and Harriers all appear to have a special fancy for feeding their young brood with this particular bird, as long as any are to be met with in their neighbourhood.

The old birds, together with their young, were obtained between Shoreham and Worthing, on the coast of Sussex, in June, 1874.

HOUSE MARTIN.—(Immature.)
Case 18.

This case. which represents the birds clinging to the face of a cliff, is copied from a sketch made under Tantallon Castle, on the coast of East Lothian.

There are several ledges of rock along the shore between Seacliff and Canty Bay where House Martins may be found nesting every year. Their nests are, however, so much the colour of the rocks, that it takes some time to discover their whereabouts.

The specimens were obtained partly in August, 1874, in East Lothian, and the remainder in Sussex, in September, 1875.

ROCK PIPIT.
Case 19.

This bird may be observed round our shores from north to south. As its name implies, it frequents cliffs and the precipitous rocks that overhang the sea on many parts of the coast.

It has a peculiar fancy for breeding on any small island in preference to the mainland. I have noticed this particularly the case at the "Ferns" and Bass Rock, together with all the islands in the Firth of Forth on the east coast, and on the numerous small patches of rock lying off the coast of Ross-shire and Sutherland on the west.

The case is copied from a sketch made on the Bass · the nest was placed among some fallen stones in the passage leading through the Fortifications. Great numbers of these birds nest on the rock, generally among the buildings, or on the ledges on the south side.

The specimens in the case were obtained at the Bass Rock, in May, 1867.

TREE PIPIT.

Case 20.

This Pipit is only a summer visitor to our shores; it may, however, be met with in most counties during the nesting season. Though possibly proceeding to the north of Scotland, I have never myself observed this bird beyond the Forest of Glenmore, in Inverness-shire, where in the summer of 1869 I found it breeding in considerable numbers.

It has a pleasing note, and is known to bird-fanciers by the name of Singing Titlark.

The specimens in the case were obtained near Brighton, in June, 1875.

SCANDINAVIAN ROCK PIPIT.

Case 21.

This bird has given rise to considerable discussion among scientific naturalists. In my humble opinion, however, it is only a northern form of our own Rock Pipit. Early in March I have shot several specimens, which plainly showed that its winter dress was identical with that bird, only a very few of the vinous feathers being visible at that time. As Spring advances, the vinous tint gradually spreads over the whole of the breast, and the back of the head and neck becomes a bluish grey.

In this plumage it may be found along the south coast from the second week in March till the latter part of April, usually frequenting the small brackish pools near the sea beach ; in some seasons, though its numbers vary considerably, it is remarkably plentiful between Brighton and Worthing. I have visited its favourite haunts on several occasions during the last three years that I have been on the south coast, but not a specimen have I met with.

In March, 1871, I shot a single bird on the Norfolk coast, near Horsey.

The specimens in the case were obtained partly at Portslade, in March 1866, and the remainder near Shoreham, in April, 1870.

TURNSTONE.— (IMMATURE.)

Case 22.

A few young Turnstones occasionally make their appearance as early as the beginning of August; they do not, however, show in any considerable number till after the gales with which we are usually visited in September and October. On their first arrival they are remarkably fearless.

Two of the specimens in the case were shot sitting at a pool of rain-water in company with a Purple Sandpiper, in the middle of the carriage-drive along the South Denes, at Yarmouth, during a gale of wind, in November, 1872; the remainder were killed earlier the same autumn on Breydon mudflats.

ROBIN
Case 23.

The case represents a pair of Robins in a woodyard, showing the block, hedging-glove, and chopper.

The specimens were obtained near Brighton, in March, 1866.

BEARDED TITMOUSE.—(SUMMER.)
Case 24.

The drainage of marshes and reclaiming of waste lands all over the country are banishing scores of our native birds from the strongholds they have held for ages.

This handsome little bird, however, unlike some of the larger species, is at present in no danger of being entirely driven from our islands, as the more extensive broads and meres in the eastern counties offer them a safe retreat. The districts, however, that are suited to their habits are fast becoming much reduced; several spots where they were formerly common in

Kent and Sussex having become completely changed by the new style of farming and other innovations.

The never-failing persecution they suffer from dealers and collectors tends also to greatly restrict their numbers.

The price of four shillings a dozen, which is offered for their eggs, induces the natives of those dreary wastes to search diligently, and but few of the first nests ever escape their sharp eyes. After the reeds get up to a certain height, it is more difficult to make out the whereabouts of the birds, and consequently the later broods escape. No one but a practised hand would ever discover the nests of this species.

There are, however, in the fen and broad districts, generally a class of men who make a living by egging, gunning, and fishing. This occupation seems to have been handed down from father to son, but I am afraid that, like many of the rarer denizens of the swamps, they will before long be either driven from their quarters, or forced to adopt a new style of life.

The specimens in the case were obtained on Heigham Sounds, in Norfolk, in May, 1870.

ROBIN.—(IMMATURE.)

Case 25.

The juvenile Robins do not assume the dress of the mature birds till after the first moult.

The specimen on the left side of the case shows a few of the red feathers already appearing on the breast.

The birds were obtained at Potter Heigham, in Norfolk, in July, 1873.

WRYNECK.

Case 26.

This is only a summer visitor, arriving early in April, and after rearing its young, leaving us before the cold weather sets in.

It is a well-known bird in most of the southern and midland counties, breeding frequently in fruit-trees in gardens in the immediate vicinity of houses.

In some parts it is known by the name of Cuckoo's Mate, its arrival being generally noticed shortly after that of the well-known harbinger of Spring. Anyone who has watched one of these birds sunning itself on the limb of a tree, and remarked the curious contortions it indulges in, can scarcely fail to understand the reason the name of Wryneck is applied to it.

The specimens in the case were obtained in the neighbourhood of Brighton, in May, 1866.

NUTHATCH.

Case 27.

This bird is generally found where large timber is abundant. It is a near relative to the Woodpeckers, and, from its somewhat similar habits, is known by that name in some districts.

It has a curious custom of plastering with mud the apertures to the holes in the trees where it breeds. A representation of a very curious nest built with mud in a haystack was given in (I believe) *The Field* some years ago. It was well authenticated.

The specimens were shot at Plumpton, in Sussex, in April, 1866.

HEDGE-SPARROW.

Case 28.

This well-known bird is represented with its young brood.

The fleshy substance round the beak of the young was carefully copied and coloured from a living "model."

The specimens in the case were obtained near Brighton, in May, 1874.

BEARDED TIT.—(IMMATURE.)

Case 29.

The case represents the young birds as soon as they are full fledged. In this early stage it is easy to distinguish the males from the females, as the former have already assumed the orange-coloured beak, which always remains one of their distinguishing points.

The present specimens were obtained in September; but young birds in this state of plumage may frequently be seen by the middle of May, or even earlier.

The Bearded Titmouse commences its nesting operations as early as the middle of March, and would in all probability rear two or three broods in a season if unmolested. The demand for eggs, however, is so great, that but few of the earlier nests escape.

The specimens in the case were shot on Hickling Broad, in Norfolk, in September, 1872.

WOOD SANDPIPER.—(SUMMER.)

Case 30.

The Wood Sandpiper is most frequently observed in the autumn. It is, however, occasionally met with in the spring and summer, and has once or twice been noticed breeding in the northern parts of the island.

In June, 1867, I found a pair of these birds on Gullane Links, in East Lothian, and from their actions I have not the slightest doubt they were breeding close at hand ; but one of them being accidentally killed, I was unable to discover their nest.

The bird shot was a female, and had evidently been sitting.

The specimens in the case were obtained on the marshes round Hickling Broad, in Norfolk, in May, 1870.

WOOD SANDPIPER.—(Autumn.)

Case 31.

The immature birds of this species used to be plentiful in the marshes near Rye, in Sussex, some years ago ; but since the drainage of the ground and other alterations, they have never appeared in such numbers. They may, however, occasionally be seen in flocks in August and September in any flat, marshy distiict, remaining sometimes for weeks, if not meeting with too warm a reception.

I have never fallen in with this bird in the winter months.

The specimens in the case were shot in the marshes between Rye and Winchelsea, in Sussex, in August, 1858.

BEARDED TIT.—(Winter.)

Case 32.

In winter the Reed Pheasants* join in flocks, varying from three or four to ten or fifteen, and keep together till early in the spring.

Though such a delicate-looking little bird, they are remarkably hardy, and seem able to contend against severe weather with greater success than many much larger and apparently stronger birds.

The specimens in the case were obtained in the reed beds round Heigham Sounds, in Norfolk, in December, 1871.

WHITETHROAT.

Case 33.

These lively little migrants soon make their arrival known, by showing themselves singing and chattering on the top of the first hedge they reach after landing on our shores for their summer visit.

They seem happy enough when they reach our coast, but I have noticed them very hard pressed during a fresh north-west wind in the Channel.

They appear to fly low to escape the force of the wind, and this unfortunately leads to their being struck down by the spray, when some unusually heavy sea happens to break right in front of them.

I believe that the smaller birds of passage seldom

---

* The Bearded Titmouse is known among the marshmen in the east of Norfolk by the name of Reed Pheasant.

attempt to cross in the face of a gale, but that they will occasionally make a mistake in the weather I have good proof, as I have picked up several (particularly of this species) floating dead on the water, a few miles off the south coast.

They are remarkably neat and handsome birds when they first arrive, but the cares of a family, together with the toil of providing for their wants, soon takes the gloss off their coats, and by the end of the summer they generally have a most ragged and disreputable appearance.

The specimens in the case were obtained in the immediate neighbourhood of Brighton during the summer of 1870. The old birds are not the parents of the young, being shot earlier in the season.

LESSER WHITETHROAT.

Case 34.

The Lesser Whitethroat is by no means so abundant as the Common. I have, however, noticed it in considerable numbers in the grass country about Harrow-on-the-Hill, in Middlesex. On the south coast it shows itself most commonly early in the autumn : it is then on the point of leaving us for the winter.

The nest of this species is a particularly light and finely-interwoven structure, being just sufficiently strong to carry the weight of the young brood, and at the same time so slightly built as almost to give the impression of being old and deserted.

The specimens in the case were taken in a clap-net close to Brighton, in August, 1869

GOLD CREST.

Case 35.

This bird is plentiful from north to south, occurring most frequently in the neighbourhood of large fir plantations. Though it remains with us as a resident throughout the year, I have noticed in several counties that fresh arrivals take place in the autumn, and such being the case I suppose a corresponding number leave in the spring.

I have never myself met with this species while crossing the North Sea during the autumnal migration, but several fishermen (who knew the bird well) have assured me that scores have occasionally settled on their boats to rest; one man in particular stating that they would roost all night in any shelter they could find, some creeping into the blocks, where, he remarked, they would remain " weeping all night," and in the morning would fly down and pick about on the corks and other portions of the nets that were out of water. As I discovered that they were known to the men by the name of Herring Spink, I was particular in inquiring whether they referred to the Chaffinch, but I found they were well acquainted with that bird also. I have, when cruising with the herring fleet in the North Sea during the autumn, generally noticed the Chaffinch as one of the most frequent visitors to the boats. Some of them appear much distressed by the journey, and after pecking about the deck for some time, often go to sleep in an old fish-basket or any quiet corner they can find.

The masters of most of the light ships off the Norfolk and Lincolnshire coasts have, for some time past, very kindly been in the habit of keeping for me the right wing of all the birds that strike the lamps; and although I have examined thousands of these wings from time to time, I never saw but one belonging to the present species.

During the autumn of 1863, I found early one morning a fir plantation, on the coast of east Lothian, almost covered with these birds; there must have been scores on every tree for several acres. How such a gathering can be accounted for I am at a loss to judge, unless they had just crossed the sea from the north of Europe. I visited the same place the next day, provided with a good supply of dust shot, being determined to try and discover if any Fire Crests were among their numbers, but the flock had entirely disappeared.

The specimens in the case were obtained in the spring of 1869, in Tarlogie Woods, near Tain, in Ross-shire.

TWITE.

Case 36.

These birds may generally be found every winter in flocks along the south coast, either by themselves or in company with Linnets and other small birds.

They seem to have a particular fancy for some food that is found among the weeds on the mudbanks that are covered by the high tides; if driven from these spots, they are sure to return after a short flight.

I have frequently, when in Caithness and Sutherland, endeavoured to discover the nest of this species, but although the birds themselves were plentiful on some parts of the moors, I never succeeded in finding either young or eggs.

Some of these birds appeared to remain in small flocks all through the summer.

The specimens in the case were shot on the mud-banks in Shoreham Harbour, in December, 1869.

QUAIL.

Case 37.

This bird is much commoner in the southern counties than is generally supposed.

Numbers breed in the immediate vicinity of Brighton, their eggs being frequently mown out within a few hundred yards of the town. Still, as but few are ever obtained during the shooting season in this part, I suppose they must be migrants on the south coast, although their time of arrival and departure, or where they go to, **has** always been a mystery to me. They are occasionally found in most counties in England, though but rarely in Scotland.

I have often myself fallen in with several scattered birds when shooting in the Fens, near Cambridge, where, with the help of Partridges, Snipe, Duck, Teal, and Hares, a good mixed bag was not unfrequently made up during the early part of the winter. I now and then heard of as many as seven or eight brace being killed in a day, but that was usually in the beginning of September. I have also good proof that they nested in the neighbourhood, as one of the specimens in the case was shot on Bottisham Fen, in May, 1861; the remainder being bagged on Swaffham Fen, near Cambridge, in November, 1862.

DIPPER.

Case 38.

There appears to be a difference of opinion among writers concerning the habits of these singular birds; some declaring that they are prejudicial to the spawn and fry of fish, while others assert that they are not only perfectly harmless, but of the greatest service to the rivers, from constantly feeding on various kinds of destructive beetles and other insects.

I believe that there is undoubted evidence that they have been both seen and shot while in possession of small fish; but I am decidedly of opinion that their presence on the waters is beneficial rather than otherwise.

They are to be found frequenting most of the rivers and streams in the northern parts of the Island, being perhaps more numerous in the rocky burns of Perthshire than in any other county.

During hard winters I have occasionally fallen in with these birds on the seacoast, being, I suppose, frozen out from their accustomed streams.

The specimens in the case were obtained on the river Lyon, in Perthshire, in November, 1865.

WILLOW WREN.

Case 39.

This small migrant is widely distributed over the British Islands, being particularly numerous even to the north of Scotland, wherever there is sufficient vegetation in the various glens.

The great similarity between this bird and the Chiff-

chaff leads many to confuse them. There is, however, a very marked difference in their song, in the
position and construction of their nests, and also in
the marking of their eggs.

I have seen it stated that a Cuckoo's egg has been
found in the nest of this species. It is, of course,
well known that the Cuckoo occasionally carries its
egg in its mouth, and is in that manner enabled to
insert it in nests that would otherwise be unapproach-
able to a bird of such size. I, however, am unable to
understand how the young Cuckoo, when arrived at
the proper age, would ever be enabled to make its
exit from its nursery; the aperture of the nest only
being constructed of sufficient size to allow of the
entry or departure of its rightful owner.

I have noticed these birds very busy in gardens in
July and August, pecking about under the leaves of
the currant bushes, and discovered that they were
making a wholesale clearance of the little green bugs
that infest the plants at that season.

The specimens in the case, together with the nest,
were obtained in Glenlyon, in Perthshire, in June,
1867.

LINNET.—(Summer.)

Case 40.

Brown Linnets do not at the present time appear to be
nearly so abundant as they were in my bird-nesting
days, about five and twenty years ago.

Improvements in agriculture, such as breaking up
of waste lands and furze-covered downs, together with
the persecution they annually suffer from the nets of

the bird-catchers, have kept on gradually reducing their numbers, till in some parts of the country they are now almost become rare birds. They are, however, still numerous during the early part of the winter near Brighton. I find the average take for one pair of nets of a morning is between thirty and fifty dozen. The hens are killed, the males sent to London.

I have noticed them some winters joining with the Twites, and frequenting the weeds that grow on the salt-water mudbanks in Shoreham harbour.

Some years ago, in East Lothian, I discovered a nest of this bird in the side of a wheat stack at about fourteen feet from the ground. The young were just on the point of flying. It was rather singular that in the thatch of the adjoining stack a partridge was sitting on fourteen eggs.

The specimens in the case, male, female, and brood, were obtained on the Downs in the neighbourhood of Brighton, in June, 1870.

LINNET.—(Winter.)

Case 41.

It will be seen by the specimens in the case that during winter the male birds of this species lose the handsome crimson feathers that adorn their breasts in summer, and take in their place others of a more sober claret colour.

If kept in confinement they never regain the bright colour when they have once lost it.

The specimens in the case were obtained between Rottingdean and Newhaven, in Sussex, during the severe weather in December, 1874.

CHIFF-CHAFF.

Case 42.

The Chiff-chaff is usually the first of the warblers to visit us in the spring. It spreads over the greater part of the country, and is said to be observed frequently in the north of Scotland.

As many writers have recorded this fact it is probably correct, but I have myself been unable to detect any in the Highlands, though the Willow Wren and Wood Wren are particularly numerous even in the wildest glens.

The eggs of these three small species of Leaf Warblers (as they are styled by some authors) are much more easily distinguished than the birds themselves.

That of the Willow Warbler is, like the others, of a pinkish white ground, strongly blotched with red spots towards the longer end. The markings on the egg of the Chiff-chaff are of a claret colour, while the egg of the Wood Wren is thickly speckled with warm brown.

The specimens in the case were obtained near Plumpton, in Sussex, in April, 1867

SPARROW HAWK.

Case 43.

I am afraid this Hawk has many enemies and but few friends. Those who take an interest in falconry often put in a plea for that still greater robber, the Peregrine ; but I never yet heard a word spoken in defence of the poor Sparrow Hawk.

That he is a bad character no one who has taken the trouble to watch his habits can deny.

Gamekeepers often wrongfully accuse certain birds of carrying off the young pheasants from the coops during the breeding season; but with regard to this hawk, I always think that he fully deserves the bad name which he has acquired.

This bird would, I should imagine, soon become a very numerous species, were it not so universally killed down. It is seldom that a brood is allowed to fly on any ground where game is preserved, except where they owe their safety to the denseness of the timber or the laziness of the gamekeepers and trappers.

The Sparrow Hawk is found all over the country, being most numerous in the neighbourhood of large woods.

It generally makes use of the nest of the Crow or some other large bird in which to rear its ravenous brood. I have noticed in the Highlands that the young birds were frequently fed upon the Meadow Pipit, which is one of the commonest small birds in that part of the country during the summer months.

The specimens in the case, both old and young, were obtained near Lairg, in Sutherland, in June, 1868.

The nest had been built and occupied by a pair of Grey Crows the previous year.

MERLIN.

Case 44.

Immature birds of this species are frequently met with in the south, though the true home of this dashing little hawk is evidently in the land of the heather and mist.

They are said to be very destructive to game, and as such usually pay the penalty that the possession of a bad name incurs. Whether it is that my own experience with regard to this bird has been too limited to form a correct judgment I am unable to say, but I hardly think that they are the desperate characters that they are generally described. Those which I have seen in the south were usually in pursuit of small birds, and while seeking this sort of prey they are frequently captured in the clap-nets that abound near Brighton.

On the Grouse moors in the north I have examined the remains of the victims that they have consumed near their nests, and never found anything larger than a Dunlin, which bird, with Larks, Pipits, and large moths (principally of the Egger species), seemed to make up their bill of fare. Though frequenting most of the wild, rocky glens in the Highlands, they seem to have a partiality for the more open moors, being particularly numerous in the flat parts of Sutherland and Caithness.

The nest is generally placed amongst the heather on the ground in the open moor. The eggs in the case were, however, taken from the face of a rock overhanging a hill loch in Ross-shire.

The female was shot, but, being a good deal injured,

was not retained ; while the male, falling winged among large stones, managed to make good his escape into some hole before I could reach the spot. While searching for him I stumbled on one of the best concealed whisky stills I ever met with. It will certainly be a particularly cute exciseman that discovers its whereabouts without the help of previous information.

The male bird in the case was trapped at a nest in Strathmore, in Caithness, in June, 1869, and the female shot in Sutherland the previous month.

KINGFISHER.

Case 45.

There is, unfortunately, a certain class of prowling gunners who never can resist a shot at this beautiful and harmless little bird ; beautiful it certainly is, though its beauty departs with its life, as the effigies one sees in the windows of the ordinary taxidermist are only a caricature of the living bird.

The Kingfisher is well known to anglers as a sociable companion on the banks of the streams they both love so well.

During the autumn I have in days gone by often noticed as many as forty or fifty of these birds fishing in the channels among the mudbanks in the Nook at Rye, in Sussex. They used to commence working down the creeks soon after the turn of the tide, and closely following the falling water, they found abundant food in the numerous shallow pools. About half-flood they used generally to make a move, flying up the creeks, and so on to the small drains that led through the marshes, and then dispersing themselves over the levels. I have occasionally observed a score or two fly past in small parties of threes and fours within a quarter of an hour while I was watching them from the shingle banks close at hand—one or two now and then steadying themselves for a moment, and then making a dash after a shrimp or small sea fish.

Fifteen years make a difference in most things; the mudbanks and creeks are certainly gone, and I expect the numerous parties of Kingfishers that frequented them remain only in the memory of those that have had the pleasure of watching them.

The last time I visited the spot, some fine specimens of Southdown mutton were grazing stolidly and complacently on the luxuriant turf that had formed where previously hundreds of acres of mudbanks were covered by every flowing tide.

I see that this habit of coming down to the salt water, and occasionally into the harbours, is common all along the south coast during the autumn.

In the Broad districts in the eastern counties the birds are not nearly so common as might be expected.

While Snipe shooting one winter round Hickling Broad, in Norfolk, I noticed some small object splashing in the water at the side of a dyke, and on proceeding to the spot I discovered an unfortunate Kingfisher, that had come to grief in a rather singular manner. The bird had evidently at some former time been struck by a shot, which had passed through the upper mandible. This wound was quite healed up, but a small piece of the horny substance of the beak had been splintered, and into the crack produced by the fracture two or three of the fine fibres which form part of the flowers or seeds of the reed were so firmly fixed, that the bird was held fast. It must have been flying up the dike, and, brushing too closely to the reeds that grew on the banks, been caught in the manner described.

The struggles of the captive had broken down the reed, which was lying flat on the water, except when lifted up by the victim in his vain attempts to escape. On being cleared from his unpleasant position he flew off, apparently none the worse for the mishap.

The specimens in the case were shot between Shoreham and Lancing, in Sussex, in January, 1871, the case itself being copied from a small sluice on the saltings in Shoreham Harbour.

SHIELDUCK.

Case 46.

These birds are abundant on many parts of the coast. They may be found as residents, however, more commonly in the northern than in the southern division of the island ; but in severe weather they are frequently driven from their usual haunts, and make their appearance on any open water.

In some quarters they are known as "Burrow Ducks," their name being derived from their habit of breeding in rabbit warrens, the nest being usually placed at a depth of four or five feet in the burrow.

By the time the young are hatched, the parent birds (the females especially) generally present a very dirty and ragged appearance ; the confined entrance and passage to their nursery most probably being the main cause of their threadbare condition.

Most waterfowl conduct their newly-hatched young under the shelter of the reeds or long grass that may be found near at hand⁎; but this species, if not out at sea, may usually be observed with their brood on the open sands.

The young birds, though seen in such a seemingly unprotected state, are by no means easily procured.

On the first signs of danger they scatter in all directions, and each one taking a line for itself, it is seldom that more than one falls a victim to the pursuer. If surprised among the sandbanks and bent grass, their colour so resembles the surrounding objects that they may almost be trodden upon without being perceived.

The male and female in the case were obtained just before the breeding season, being shot early in the spring of 1867, in Gullane Bay, in the Firth of Forth. The old birds belonging to the brood were killed, but the soiled condition of their feathers would only have given a very poor idea of what handsome birds they had been a few weeks earlier. The young were taken in the Dornoch Firth, in June, 1868.

SHORT EARED OWL.

Case 47.

These birds are permanent residents in many parts of the island, but numbers of fresh arrivals make their appearance in the autumn, usually showing themselves about the same time as the first flights of Woodcocks in the eastern counties; and this fact, together with their somewhat similar flight, has led to their being called in some districts the " Woodcock Owl."

This species appears capable of taking its prey by day as well as night. I have often noticed several of these birds hunting over the marshes in Norfolk while the sun was up, and during dull weather they all seem out in search of food by 3 p.m.

They confine themselves entirely to the ground, breeding amongst the heather on the open moors of Scotland, and in the southern counties frequenting the rush marshes and other waste lands.

On their first arrival in the autumn they are not unfrequently flushed from turnip fields, occasionally being found in such situations in considerable numbers.

The specimens in the case were obtained in the Hickling Marshes, in Norfolk, in December, 1871.

KESTREL.

Case 48.

This is by far the most numerous of the Hawk tribe in Great Britain. Though not generally so regarded, it is one of our most useful birds, being a decided ally both to the farmer and game preserver. I have been so frequently assured that Kestrels have been detected preying upon young game, that I suppose some misguided old bird must, when greatly pressed by the cares of providing for a hungry brood, have snatched some precocious young pheasant from the neighbourhood of the coops, and, like many another poor bungling thief, been caught at the first attempt, while the greater rogues go free. The rats alone that these birds destroy while procuring food for their young would commit ten times more damage in one year than the poor inoffensive Kestrels could possibly effect in their whole lives.

This bird breeds in a variety of situations. In the

south it may be found nesting in chalk pits and in the cliffs overhanging the sea. In the midland counties it will make use of any old deserted nest when other accommodation cannot be found; and on the moors in the north the steep rocky faces of the numerous ravines and old ruined buildings are for the most part resorted to. In the summer of 1868 I found a brood of young Kestrels on a perfectly bare spot on an open moor in Sutherland, and in the following year I was shown three eggs on a shelf in a shepherd's shealing on Ben Alisky, in Caithness, which had only been deserted by its rightful owners the previous month · the bird making its escape through a hole in the chimney while we were entering the building.

The specimens in the case were obtained along the coast near Canty Bay, in East Lothian, in June, 1867. The case itself is copied from a Kestrel's nest in a rock overhanging the road between Kenlochewe and Gairloch, in Ross-shire.

LONG-EARED OWL.

Case 49.

This species is common in most parts of the country, being found, however, more plentiful wherever there are fir plantations of any size.

Unlike the short-eared Owl, these birds are seldom seen by day. Soon after sunset they leave the shelter of the woods, and search the adjacent fields and hedge rows for their prey.

When I lived in East Lothian, I used to observe these Owls during the summer, coming regularly at dusk to the stacks for rats and mice, though the woods where they nested were at a distance of nearly two miles.

The young birds have a particularly sad and plaintive whistle (something resembling a deep-drawn sigh), when calling for their food. Where there are several broods in the same plantation, the effect of their wailing cries is anything but lively, when listened to on a still night in the gloomy depths of the pine woods; the mournful notes breaking out first on one side, then on another, and finally being answered from all quarters at once.

This species occasionally preys upon young birds. I one evening noticed a Long-Eared Owl making several visits to a boat-shed on one of the broads in Norfolk, and on examining the place the next morning, I discovered that a brood of young Swallows had disappeared during the night.

The specimens in the case were obtained in the neighbourhood of Norwich, in June, 1871.

CURLEW.

Case 50.

During the summer months these birds resort to the hills and moors, where, in company with the Red Grouse and Golden Plover, they rear their broods. In the autumn they may be found in flocks, sometimes feeding on the coast, and occasionally on the stubbles and turnip fields. On the approach of winter they generally take up their quarters on the mudbanks in some tidal harbour, or on any extensive flats along the coast, where they are free from persecution. If not frequently disturbed, they are by no means shy, but after a few shots they become one of the most wary of birds, their well-known cry serving as a signal of danger to all the wildfowl within hearing distance.

They are a first-rate bird for the table, particularly when a few severe frosts have taken down a little of the fat with which they are almost covered when they first arrive on the mudbanks.

I have once or twice observed small flocks of young birds in the Nook, at Rye Harbour, in Sussex, as early as July; but this I should imagine must be two or three months sooner than they are usually found so far south.

As will be seen by the specimens in the case, the young when first hatched have only a short bill like a Plover.

The old birds with their brood were obtained on the hills in Glenlyon, in Perthshire, in June, 1867.

WILD DUCK.

Case 51.

Though not so numerous as some species of our British wildfowl, the Wild Duck is by far the most generally known, occurring at various seasons in all parts of the island.

There have been several discussions in the sporting papers about the time that this bird commences nesting, some of the dates given being a month or two in advance of others.

There can, however, be little doubt that in some localities the birds pair and the eggs are laid considerably earlier than in others.

Local naturalists frequently fancy that the habits of the birds all over the British Islands must be the same as they have observed in their own particular district.

The Wild Duck and Wigeon are perhaps the best flavoured, and consequently the most sought after, of all our wildfowl, the only others that can compete with them being the Pintail, Teal, and Pochard.

Even if they frequent the sea during the day, they seldom acquire the fishy taste that is so common in some fowl. Towards night they are sure to make their way inland to the marshes and rivers for food. I often noticed, when in the south of Scotland, that these birds resorted to the potato fields in large flocks, remaining all night scattered over the ground, searching for the exposed roots, and leaving just before daybreak for the open sea.

The male bird, though represented in the case as sunning himself on the bank in the society of his wife and family, is in reality but little given to a domestic life, usually leaving the brood to the care of the female, and seeking his own pleasures in company with two or three faithless husbands like himself.

The female and young were obtained in Glenlyon, in Perthshire, in June, 1867, the male being killed earlier in the season.

WIGEON.

Case 52.

Though it has been stated that this bird but rarely remains to breed in the British Islands, considerable numbers usually rear their young in three or four of the most northern counties of Scotland. When search-ing for the nests of other species, I have on one or two occasions, stumbled by chance over ten or fifteen nests in one day, and this alone will show that the Wigeon remains with us in sufficient numbers to be styled a " resident."

By the beginning of October, immense flocks make their appearance on the north-east coast of Scotland : these are probably from the north of Europe. On their first arrival they are quite unsuspicious of danger, and hundreds fall victims to the numerous punt gunners, as many as forty, fifty, and occasionally sixty, being bagged at one discharge of the big gun. The warm reception they meet with soon drives the greater part of the birds south, and during the depth of the winter but few remain in the north. At this season they may be met with all round the coast, and on any large sheets of inland water that remain open.

The young, in the downy state, have the same markings as the young of the Wild Duck, but the ground colour is a much warmer brown.

The male and female were killed in the Dornoch Firth, on the north-east coast of Scotland, in April, 1869, and the nest and eggs were taken in Strathmore, in Caithness, the following month.

OYSTER CATCHER.

Case 53.

The " Sea Pie " as this bird is sometimes called in the south, is only an occasional visitor to Sussex and the adjoining counties, but in the Channel Islands, and again in Scotland, it may be found at all seasons as a resident.

The Scotch name of "Mussel Pecker" seems much more appropriate to this bird than that of Oyster Catcher. It might possibly have a chance to make a meal off a few mussels by swallowing some of the smaller shells whole, like an Eider or a Herring Gull, but what use an oyster could possibly be to it, I am at a loss to understand.*

These birds occasionally collect in immense flocks · I have frequently found at least two or three thousand feeding together on the mudbanks at the Little Ferry, near Golspie, in Sutherland. From never having been disturbed they were perfectly fearless, and would not take the slightest notice of a boat, allowing us to scull past in the punt within five or tén yards, while in pursuit of other fowl. Being of little or no use, I never tried a shot at them with the big gun, though, had I been so inclined, I might easily have bagged from 150 to 200 at a shot.

The Oyster Catcher chooses a variety of situations for breeding purposes. At the Fern Islands, it lays its eggs on the shingle and sand at a short distance above high-water mark. On several of the Scotch rivers, such as the Spey or the Tay, it nests amongst the rough stones that form the banks of the river, while in some districts in the Highlands it chooses any open spot in either a potato or oat field, where, until the crops get up, it sits plainly in view of everyone that passes within a mile of the spot. I have also seen its eggs on some of the large detached rocks that

---

* I have been informed that it frequently crushes the shells of the mussels, and, extracting the fish, leaves the fragments scattered over the rocks, as an irresistible proof of the strength of its beak.

are found off the west coast of Ross and Sutherland.

The specimens in the case were obtained on the
banks of the Lyon, in Perthshire, in June, 1867.

BARN OWL.

Case 54.

This useful bird is so generally distributed, and so
universally well known, that any remarks I could
make on its habits would be superfluous.

The male, female and brood, were obtained in the
neighbourhood of Brighton, in June, 1872.

The case is copied from the bell tower of Chiltington
Church, near Plumpton, in Sussex.

TAWNY OWL.

Case 55.

Though by no means an uncommon, this is rather a
local species.

As its name of Wood Owl denotes, it frequents those
parts of the country that are most densely timbered.

It usually nests in a hole in a hollow tree, or in the
deserted nest of some other bird, but has in a few
instances been discovered breeding in a rabbit burrow.

Like all the rest of the family, it is a most useful
bird, but unfortunately does not generally meet with
the protection it so well deserves.

The specimens in the case were obtained near
Balcombe, in Sussex, in June, 1875.

BLACK CROW.

Case 56.

This bird may be found generally dispersed over the country, though nowhere very abundant. The bad character which it bears, and the persecution it undergoes in consequence at the hands of the game-keepers, easily accounts for its numbers being kept down.

I have never noticed these birds to flock together like Grey Crows, but the brood of the previous summer seems to keep with the parents during the whole of the autumn and winter; the family apparently only breaking up on the approach of spring or the death of some of their number.

The bare space at the base of the bill of the mature Rook is always supposed to distinguish that bird from the present species; but as old Rooks occasionally retain the black feathers above the beak, it is as well to know that the colour of the mouth of a Rook is a dull slate, while that of a Crow is a pale flesh.

The nestlings of both species show this difference as well as the adults.

I have frequently observed this bird pairing with the Grey Crow in the Highlands, and I believe it is generally supposed when this is the case that the young always take after one or other of the parents.

In the summer of 1866, when living in the north-west of Perthshire, I trapped one young bird and shot another near the same spot, that had every appearance of being a regular cross between the two species: the whole of the body was black, except a small patch of grey on the neck and back.

I at first thought they might possibly be young Jackdaws, not being quite certain whether that species had a white eye in its immature state; but, on examining a Jackdaw's nest, I soon discovered that the iris is the same colour in the mature and immature birds.

The specimens were unfortunately not preserved, as the weather was so hot that they were spoiled before I was able to send them away.

A few days later two more young birds in similar plumage were killed in the same glen by the keeper, who said that he saw them flying after a pair of old Crows, one of which was black and the other grey.

The specimens in the case were shot in the marshes near Hickling Broad, in Norfolk, in January, 1873.

They were disturbed in the act of making a meal off a fowl that had escaped wounded from some of the gunners.

PINTAIL.

Case 57.

This handsome duck is more common in the northern parts of the island; a few, however, generally show themselves during the winter all round the coast, while in unusually severe weather I have found them plentiful in both Norfolk and Sussex.

On the north-east coast of Scotland these birds are known to the local gunners by the name of "Wigeon Leaders," their greater size and length of neck always making them the most prominent birds in the flocks of Wigeon with which they are generally found associated in the firths on the coast of Ross and Cromarty.   They are also by far the most wary, and always rise first on the near approach of danger.

I have seen a few immature Pintails, which showed some of the down on the head, killed in this country. They seemed almost too young to have crossed from the Continent, but that, I suppose must have been the case, as I have never heard of this duck nesting in Great Britain.

The specimens in the case were shot on Loch Slyn, near Tain, in the east of Ross-shire, in March, 1869.

TEAL.

Case 58.

The Teal is one of our commonest ducks, numbers remaining with us all the year round to rear their young, and large flocks arriving from the Continent in the winter, generally making their appearance a day or two before we are visited by severe weather.

They are usually unsuspicious of danger, but, like all wildfowl, they have their restless moods, and when this is the case, it is almost impossible to approach within gunshot of them.

The female and brood were obtained on a hill loch, in the west of Ross-shire, in May, 1868; the male being killed, near Bonner Bridge, earlier in the season.

NORFOLK PLOVER.

Case 59.

The true home of this bird in the British Islands is, as its name denotes, in the eastern counties. On the large warrens in the neighbourhood of Thetford and other parts of Norfolk it breeds abundantly.

On the range of the South Downs in Sussex, from above Worthing to Newhaven, it is also by no means scarce, being perhaps most plentiful on the hills between Brighton and Lewes.

I noticed that all the nests I have discovered in Sussex, have been placed on slopes of the downs that faced either south or west.

I have never met with this bird during the winter, though I have heard of their being occasionally flushed from the turnip-fields late in the autumn, and it is most probable that they leave the country on the approach of cold weather.*

I believe it has been stated by most naturalists that the male and female are alike. This is certainly true as regards the plumage, but, as will be seen by the specimens in the case, the male has a knob of about the size of two peas on the base of the beak, which easily distinguishes him from his mate. We are likewise informed that incubation lasts sixteen or seventeen days, but I am afraid that the patience of the birds will have to be taxed for about five days longer, before their downy progeny breaks the shell.

The male, female, and eggs are from the hills between Brighton and Lewes, and were obtained in June, 1872.

The case is a correct representation of a nest found

---

* I was previously unaware that this Plover was found in Sussex during the winter; but to-day (January 25th, 1876) I was informed by a shepherd near Brighton, who knows the bird well, that he had just seen five flying together.

near Falmer, every stone and stem of furze being brought from the identical spot.

GREY CROW.

Case 60.

Though only a winter visitor to the southern parts of the island, this bird may be found at all seasons in the Highlands of Scotland.

In the south it usually frequents the sea-coast, living on the dead fish, or any decomposing remains that may be cast up by the tide. I have often noticed Grey Crows on the large broads in the east of Norfolk, flying one after another over the litter that was washed up on the lee shore, hunting for any dead or wounded fowl that might have escaped from the gunners.

As soon as a prize was discovered, the croaking and screaming of those near at hand would soon bring the whole of the black fraternity together, and, living or dead, the unfortunate victim would speedily have its flesh torn from its bones.

In the winter of 1868 I was punt-gunning on Loch Slyn, in Ross-shire, and having made a successful shot at a large flock of Mallard, as they rose from a rough bank, I was unable, owing to the long heather that grew near the shores of the loch, to collect the whole of the cripples, as some of the wounded birds crawled into the thick cover before I could gather up those nearest at hand. About an hour later, when on the far side of the loch, I noticed several parties of Grey Crows, numbering in all at least forty or fifty birds, flying and quarrelling near the spot where I had fired the shot, and on again searching the ground, I found they had discovered and dragged from their hiding-places seven more ducks : four were picked nearly clean, but the remaining three, though quite dead, were only slightly torn.   On looking over the spot on the following morning I found two more skeletons, which I had missed on my previous search.

On the moors in the north they are, without exception, the worst vermin that a game preserver has to contend with.   They may be seen in the spring quartering the ground like setters, and the nest of a Grouse or other game bird once discovered is soon robbed of its contents.

They usually have some elevated spot to which they carry the eggs before sucking them, leaving the empty shells lying about in dozens, as if to draw attention to their bad deeds.

They are generally shy, wary birds, seeming instinctively to know when anyone is in pursuit of them.   I have often, however, shot them by driving or riding along the hill roads in the Highlands, as they take but little notice of a conveyance.

During the autumnal migration I have occasionally met with them in the North Sea, apparently tired out by their long flight, and glad of a rest on any beat or vessel they might meet with on their course.

Two of these birds and a Jackdaw, which had followed us one day in a thick fog for a considerable distance, at last settled on one of the paddle-boxes of the steamboat.   A shot or two which I fired at some Gannets at first greatly alarmed them, and one of the crows beat a speedy retreat ; it soon however, returned, and after a time they got used to the noise of the shooting and the shouts of the men who were fishing, and stalked gravely about on the bridge, seeming to take particular notice of what was going on. Towards dusk the wind freshened and the pitching of the steam-boat appeared to disagree with them, as after looking very miserable for some time in their vain attempts to keep their footing, they at last took a reluctant farewell, flying slowly against a head wind towards the land.

The male and female, together with the nest, were obtained near Lairg, in Sutherland, in June, 1869.

The case is copied from a sketch taken from a nest in the rocks at Longa Island, off the west coast of Ross-shire.

PEREGRINE.

Case 61.

Under the heading of " Hawks and the Moors," the Peregrine has given rise to many discussions in the sporting papers ; some writers declaring that they do but little damage on a Grouse moor, being of opinion that the few birds they take are usually the diseased and weakly, while others class them amongst the very worst of thieves.

These discussions evidently being, for the most part, between falconers on the one side and game-preservers on the other, there can be but little doubt that each party takes a rather one-sided view of the case.

The Peregrine accommodates itself to the district it breeds in, preying on Grouse, Plover, Ducks, and Pigeons, in the Highlands ; sea birds, such as the smaller Gulls and Guillemots, on the islands; and Partridges, Pigeons, and even the young of the domestic fowl, in the south. It is, however, impossible for any-one to give the true " bill of fare" of a Peregrine, as, on examining a dozen nests, there will be found the remains of different victims in each.

Though certainly an enemy to the game preserver in the north (every Falcon destroying on an average at least one brace of Grouse or other game birds in a day), I should be sorry to see this dashing Hawk " improved" off the face of the country, and would say, in the words of Mr. Jorrocks, " Be to his wirtues ever kind ; be to his faults a leetle blind."

The specimens in the case were obtained on the hills above the Lochs of Roro, in the north-west of Perth-shire, in June, 1867.

The birds were so wary, and the position of the nest so exposed in the bare face of a precipice of at least one hundred feet in height, that it was impossible to get a shot at either of them ; so removing three of the young I stumped the remaining one down, and set a couple of traps on each side of it.

The female was taken the same evening, but it was three days before the male bird was seen ; when, on examining the traps early in the morning, we found an unfledged duckling in the first trap and the Falcon in the second. He had evidently sprung the first trap with the prey he had been bringing, and then in his vain attempt to drag the duckling, trap and all, to the young one, had been himself caught in the second.

During the three days between the capture of the parents, we had kept the young Hawk alive by feeding it with trout, fresh caught from the Loch at the foot of the hill, on which it seemed to thrive well.

HEN HARRIER.

Case 62.

This bird (which is considered to be a connecting link between the Hawks and Owls) is common on the flat moors in the centre and east of Sutherland, and also over the greater part of Caithness.

It may, in like manner, be met with scattered over the country wherever there are large open heaths or furze-covered downs

On two or three occasions I have found its nest in the rushy marshes in the neighbourhood of the broads in the east of Norfolk.

Immature birds, the same as with other species, seem to wander more readily from their usual haunts, and are not unfrequently obtained in Sussex and the adjoining counties.

It is said to be very destructive to game, but my own observations would lead me to believe that it preys more on small vermin and birds of about the size of the Titlark than on anything larger.

In the summer of 1869, while walking over a moor in the east of Sutherland, I disturbed a Ringtail (the female of the Hen Harrier is known by this name) from her nest, which contained one young one just out of the shell, and five eggs on the point of hatching. As both the old birds were flying round in a state of great consternation, I sat down to watch their actions for a few minutes.

On rising to leave the spot I discovered I had laid my gun on the back of an old Grey Hen, who now got up from her nest, in which were three fresh-laid eggs, evidently showing that she herself had chosen this apparently dangerous locality for her nursery, as the Harrier's nest was within six or seven paces. This is not the only instance I have met with of game and birds of prey being found in close proximity.

As will be seen by the specimens in the case, the eye of the male is a bright yellow; of the female a warm brown; and in the young a pale blue.

When first I found this nest it contained five eggs, but on visiting it a fortnight later, there was only a single young bird; either the eggs or young having been carried off by some Grey Crows, which were breeding in a steep rock at no great distance.

The whole family were obtained on a moor in the west of Caithness, in June, 1868.

MARSH HARRIER.—(Immature.)

Case 63.

I have never to my knowledge seen the adult Marsh Harrier in a wild state.

Immature birds may, however, be observed commonly in the autumn hunting over the rough marshes and reed-beds that surround the broads in Norfolk and Suffolk.

They appear to prey upon small reptiles and wounded fowl that have escaped from the gun, and crawled in among the rushes to die.

The specimen in the case was shot in the Potter Heigham marshes in the east of Norfolk, in the autumn of 1871.

SPOONBILL.

Case 64.

Most old gunners can remember the time when flocks of these birds were common every spring in the marshes and on the mudbanks round our coasts. A few, however, still make their appearance nearly every season about the middle of May, along the flat country between the mouths of the Humber and the Thames.

A Spoonbill, when pitched by itself on a mudbank where food is plentiful, is generally easily approached within gunshot; but its unusual appearance seems to so excite any Gulls that are near at hand, that they immediately commence flying and screaming round the stranger, and never cease their clamour till they have driven it out of their sight.

The male bird in the case had frequented Breydon mudflats for a week or ten days, feeding, whenever he could get a chance, but had been so persecuted, that he never had time to settle for more than a few minutes before he was compelled to quit the spot.

It was only by waiting near his accustomed feeding-ground, just at daybreak, that I was enabled to get a shot at him.

It is stated that many years ago they bred in Norfolk, nesting on the tops of trees in the same manner as the Heron.

The specimens in the case were both shot on Breydon mudflats, near Yarmouth—the female in May, 1871, the male in May, 1873.

BUZZARD.—(Immature.)

Case 65.

The young Buzzards, as soon as they leave the care of the parent birds, are remarkably unsuspicious of danger, and are nearly certain to fall victims to the first trap that comes in their way.

When living in the west of Perthshire, I noticed, one autumn during the early part of September, two or three of these birds frequenting the face of a steep hill; and, setting a trap on a cairn* built up for the purpose, I took within twenty-four hours three young Buzzards, a Cat, and a Stoat.

The specimens in the case, which are two of the above-mentioned birds, were taken in Glenlyon, in Perthshire, in September, 1865.

BUZZARD.—(Mature.)

Case 66.

The Buzzard is always a lazy, indolent bird, seldom striking any prey for itself which requires more exertion to capture than a half-grown rabbit or hare; usually preferring to feed on wounded game, or those that are diseased and weakly.

When seen at a distance on the wing it bears a striking resemblance to the Eagle in miniature.

It was formerly much more abundant; but since the rage for game-preserving on an extensive scale has set in, its sluggish habits and manner of feeding has

---

* A pile of loose stones.

rendered it an easy victim to the trapper, and it is gradually becoming a scarce bird.

The specimens, with their nest and eggs, were obtained among the hills near Kenlochewe, in the west of Ross-shire, in May, 1868.

HERON

Case 67.

Formerly the head of the Game list, the Heron, since the decline of hawking, has fallen from its proud estate, and at the present day is but little esteemed by any save plumassiers.

It usually nests in trees, in smaller or larger communities, known by the name of " Heronries."

The sketch from which the case is copied was taken at the *C*romarty Rocks, on the north-east coast of Scotland, where some hundreds of these birds construct their nests in the ivy-covered face of the cliffs.

I am not acquainted with more than one or two other localities in the British Islands where Herons build in similar situations.

A few pairs, however, generally rear their young on a steep hill-side above an almost inaccessible loch in the Western Highlands.

The specimens in the case were obtained at the Cairn Rhui, on the north part of the Cromarty Rocks, in May, 1869.

BLACK GROUSE.

Case 68.

Though a few of these fine birds are still to be met with in even the most southern counties, we must cross the Tweed before we can observe them in their true home.

During August the young Black Game are usually so tame as to afford little or no sport. By November, however, when they have gained both strength and experience, it will, on most moors, need hard work to fill a bag without having recourse to driving.

I have occasionally seen them during a severe snow-storm at the end of the season so cut up by the weather, that they would sit huddled up in the birch trees, and allow themselves to be shot on their perches, if anyone were inclined to take such an un-sportsmanlike advantage of them.

On clear still mornings, during the latter part of the winter and early spring, the oldest and finest birds usually collect on some open spot, just after daybreak, and go through a regular performance ; but whether it be fighting or playing I am unable to say.

I have frequently watched them from a distance ; but on one occasion, having noticed that they had for a day or two held their meetings on the brow of a hill where I could get a good hiding-place, I resolved to make an attempt to witness the whole of their pro-ceedings.

Accordingly I arrived at the spot an hour before daybreak, and, creeping into a regular nest of rugs and plaids which the keepers arranged for me, I was covered well over with dry heather and brakes, and finally sprinkled with snow.

I then sent the men away, and quietly awaited the performance.

Almost simultaneously with the first streak of light in the east, I heard a rush of wings; and an old cock, passing within a few feet of my head, settled on the open space about twenty yards in front of me.

For full ten minutes there were no other arrivals, and I began to fear that another spot had been chosen for that day's amusements.

Suddenly three or four more appeared on the scene, having probably quietly alighted on the other side of the brae.

For a few moments they remained silently watching one another, apparently waiting for a signal from the leader.

I next caught sight of two or three small parties flying high in the air direct from the hills on the opposite side of a steep burn. After circling once round the spot, they alighted lower down on the hill, and some of them, principally grey hens, remained where they were, while the males gradually ascended the rising ground, picking their way with the greatest care, carrying their tails high over their backs, either to show themselves to the greatest advantage, or to avoid contact with the frost and snow.

These were speedily followed by others, and they kept on gathering, till between thirty and forty were collected in a kind of irregular circle.

The old cock, who had first appeared, and who was evidently looked upon as the master of the ceremonies, now advanced into the centre of the arena; his comb was elevated, his wings drooped, his tail curled over his back, and every feather—even down to his toes—was spread to its fullest extent. After bowing all round, and apparently being satisfied that no one wished to dispute his title to be considered the greatest swell present, he proceeded to execute a kind of *pas seul*, which seemed to consist of a double shuffle, hop, skip, and a jump, and was concluded by an almost complete somersault. Four others then advanced towards the open ground, two coming from either side. These went through something like a set figure; advanced, bowed, turned round, jumped over one another's heads, turned round, bowed again, and then retired.

Several more then joined in the performances, and the proceedings were brought to a satisfactory termination by the whole of the actors advancing, bowing, passing one another, turning round, bowing again, and then separating.

After this they broke up into small parties, and dispersed themselves over the ground.

There was little or no real fighting; but this may possibly be accounted for, as it was only about the middle of December, and I believe the fiercest battles are usually stated to take place in the spring.

I have occasionally noticed gatherings on a smaller scale in the evening.

The specimens in the case were shot on the Innerwick Moors, in Glenlyon, in Perthshire, in December, 1867.

BITTERN.

Case 69.

A few of these birds still visit us during the winter, generally arriving with a frost and easterly wind from the coast of Holland.

Though formerly breeding abundantly in the extensive reed-beds and swamps that were in those days common in our eastern counties, it is now some years since a nest has been discovered in any part of our island; the last authenticated eggs, I believe, being taken at Upton *C*ar, near Acle, in Norfolk.

The greater facilities for the drainage of the marshes since the introduction of steam water-mills have, together with the rage for reclaiming waste lands, gradually assisted to restrict their haunts, and thereby rendered those that visit our shores more accessible to the gunners, who are always on the track of any storm-driven stranger that makes his appearance.

Three of these birds frequented the neighbourhood of Hickling Broad and Heigham Sounds for about a week in July, 1873. They were several times put up by the marsh-men while going to and returning from their work, but managed to evade all dangers, and leave the country of their own free will.

The specimen in the case was shot in one of the reed-beds surrounding Hickling Broad, in Norfolk, in December, 1871.

STORK.

Case 70.

I suppose it is best to tell the truth at once, and confess that my knowledge of the above species (from personal observation) is confined to the single individual in the case. This bird had, I believe, been noticed for some days in Suffolk before he made his appearance in Norfolk.

I first received word of his arrival from a carrier, who, while on the road from Yarmouth to Hickling, observed him fly in from the sea, and pitch in the marshes near the coast. Here he was speedily discovered by some Peewits and Rooks, and, after continued buffetings, driven further inland. On searching the ground on the following day, I met with no success. A week later, however, I saw him rise from a marsh at Potter Heigham, and attack a Heron that was attempting to settle near his quarters. As he pitched in a reed-bed close to a dike, I had not the slightest difficulty in approaching within gun-shot.

Although he had been (as I afterwards learned) for a couple of days in a country abounding with frogs and other suitable food, there was nothing except a few large spiders in the stomach.

The Hickling keeper, who had shot one about thirty years previously, informed me that it had been feeding voraciously on young pike, which it had captured on some flooded marshes.

The specimen was shot on Rush Hills, near Potter Heigham, in Norfolk, in June, 1873.

RED GROUSE.—(Winter.)

Case 71.

This case is exhibited to show the plumage of the old male Grouse during winter and early spring.

Some specimens are much more strongly marked with white than others.

The birds were shot in Glenlyon, in Perthshire, during the winter of 1865.

CAPERCAILLIE.

Case 72.

Though thoroughly naturalized by a residence of many years, the present stock of this magnificent Grouse is only an importation from the North of Europe. Strath Glass and the adjacent Glens are said to have been the last strongholds of the native breed; and to those who are acquainted with the immense fir woods that cover the sidoo of some of the rugged hills in these localities the cause of their disappearance must for ever be a mystery.

The specimens were shot in Perthshire, in 1878.

CORMORANT.—(Mature.)

Case 73.

These birds are common round many parts of the coast, and occasionally make their appearance on inland waters.

They seem to have a partiality for resting on elevated spots, such as detached rocks or beacons at sea, stakes and posts that mark the channel in muddy rivers or fresh-water lakes, and dead trees.

I noticed an immature bird of this species settle on the gilt cock that formed the vane on the top of the Town Hall at Tain, in Ross-shire, a few minutes before seven, one evening in September, 1869. This was a most unpleasantly shaped perch, and the bird had the greatest difficulty in steadying itself; the clock striking the hour of seven disturbed it for a few minutes, but, returning again, it managed, after two or three attempts, to regain its former position. Here it remained, evidently very uncomfortable, till fired at from the centre of the High-street, and put to flight.

I have, early in the spring, met with a few of these birds with perfectly milk-white necks, but (though nearly losing the punt on one occasion in a vain attempt) I never succeeded in obtaining a specimen in that state of plumage. The white feathers must, I should imagine, be either shed or change colour before the breeding season, as I never noticed Cormorants with their necks marked in this manner at any of their nesting-stations.

They breed in the face of high rocks and cliffs, and at times on low islands, where their nests are only elevated a few feet above high-water mark. Among the sticks and other litter which they make use of for building, I have seen children's whips and spades, a gentleman's light cane, and part of the handle of a parasol, all of which I suppose the birds had picked up floating at sea.

The specimens in the case were shot at the rocks under Sneaton Castle, a few miles north of Whitby, in Yorkshire, in May, 1862.

The nest and eggs were taken at the Fern Islands, off the coast of Northumberland, in June, 1867.

PTARMIGAN.—(SUMMER.)

Case 74.

Although it may possibly be regarded as improper to exhibit a game bird shot during the breeding season, I hope that my attempt to show the Ptarmigan in its nesting plumage, and so illustrate the three seasons of summer, autumn, and winter, will be considered sufficient excuse for such an unsportsmanlike performance.

It is only among the mists near the summits of the highest hills that its nest is to be found. Here, without a neighbour, save the Dotterel, Snow Bunting, or Blue Hare, it passes the summer, till driven by the storms to seek shelter from the winter blasts in the more sheltered corries at a lower elevation.

It would soon become more numerous were it not for the tribute it is forced to pay to the mountain Fox and Raven.

The eggs of this bird are by no means easily discovered. Though frequently searched for, I never had the luck to meet with a nest except by accident. At last, after many unsuccessful attempts, three nests were discovered within a few hundred yards of one another, on the hills above Glenlyon, in the north-west of Perthshire.

I had so many times gone over the ground within eight or ten miles of the Lodge without success, that I at length determined to search the land belonging to some adjoining shootings over which I had liberty to hunt for any specimens I might require.

Starting before daybreak, accompanied by one keeper, and a gillie leading a pony with provisions and plaids, in case we did not get back that night, I had, by mid-day, gone over several of the rough hills that lay between Loch Rannoch and the Lyon ; and after about ten hours' work without having started a single female, as heavy thunder was rumbling away among the hills to the west, and there appeared every indication of an approaching tempest, I had come to the conclusion that it would be the wisest plan to return home, and renew our search in more favourable weather.

While resting for a short time after lunch on the top of the hill, before turning back, we were surprised by a shot, and on looking with the glasses far below us, we could make out three men with about a dozen dogs, trying to bolt a fox from a rough cairn of stones. We afterwards learned that although the foxes had been there very lately, as was clearly indicated by the remains of some fresh Grouse and Hares, which the terriers dragged from the earth, none of the family were then at home.

The shot had been fired at a young Raven, which had been hatched in the rocks above the fox cairn. The old birds were very noisy, but being too good judges of distance to venture within shot, unfortunately escaped.

Just then a fine cock Ptarmigan appeared on a large rock close by, and as he resolutely refused to leave the spot, running only a yard or two in advance of us when we approached him, we searched every inch of the ground, which was almost a mass of large stones, but without putting up the female, although by the actions of the male we were nearly certain she was close at hand. I also tried a brace of steady setters, which I had brought up as an experiment, though I had not much faith that they would be of any service.

On returning to the lunching place, we discovered the three men (who we soon recognised as two keepers and a shepherd) coming in our direction.

While they were making their way up-hill, the rain, which had been threatening some time, came down the thunder becoming more distant; the afternoon turned cold, a dense mist coming up with the wind.

On reaching us they were agreeably surprised to discover who we were, more particularly as they had neither meat nor drink with them, and needed but little pressing to commence operations on what we were able to provide them with.

When they had satisfied their hunger, as I found they were anxious to have the assistance of an additional gun whilst trying another large cairn, where they expected to find the cubs they had missed lower down the hill, I agreed to finish the day with them.

As the top of the hill was nothing but rocks and stones piled one on the top of the other, we were forced to leave the pony where we had lunched. Our course was now kept north, as the earth we were going to was on the Rannoch side of the hill. One hollow which we passed through seemed alive with Ptarmigan, the cocks were flying and croaking in all directions, but as no hens were started, and the weather had turned thick and dark, we resolved to leave our search for nests till a finer day.

On arriving at the fox cairn, we found it bore no signs of having been used this season, and as it was now too late to try further, the terriers were coupled and we turned back.

We had not gone more than one hundred yards when, hearing a scuffle behind us, we turned round, and saw a hen Ptarmigan struggling in the mouth of one of the fox-terriers; on shouting to the dog the bird flew away, none the worse excepting the loss of a few tail feathers.

As we expected, the terrier had seized her on her nest, which contained seven eggs. He had, luckily, only been able to catch hold of her tail, as the other terrier to which he was coupled was tugging in the opposite direction, and had most probably so saved the life of the bird. As I required the female in the present state of plumage, we determined to retire to the shelter of a large rock about two hundred yards off, and await the return of the bird to her nest.

I then sent off one of the men to the pony for what was left of our eatables and drinkables, and we made ourselves as comfortable as the circumstances would permit.

The head keeper, who had joined us, said he had observed our pony from the lower cairn, and thinking we were from the Rannoch side of the hill, he had come up to order us off the ground, for he had a great antipathy to the people who marched with him in that direction, as they frequently came on his side of the hill and shot his hares and Ptarmigan. On the last occasion he had met with them about half a mile over the march, and after informing them that he had no desire to give them another day's shooting, had sternly walked away, refusing all manner of tempting liquors which had been pressed upon him.

Poor old Sandy is gone, and I should be sorry to throw doubt on any of his statements, but still I can scarcely credit the latter part of his story, as Sandy dearly loved his native mountain dew.

We had been rather more than an hour sitting under the shelter of the rock with all the dogs gathered round us, when a colly, which had been lying within a yard of my feet, got up, shook the wet from his coat, and lay down again, this time changing his position by about a foot. He now chose the brown back of a female Ptarmigan to recline upon, which, causing a great flutter, startled the dog as much as the bird herself.

On her flying off we discovered she had been sitting on eight eggs. The nest was within a couple of yards of the spot where we had sat for more than an hour, and it was a wonder, with so many dogs about, that some of them had not stumbled on her sooner. As this bird would be the most perfect, I determined to obtain her instead of the one which had been caught by the terrier.

It was now getting late, so, after a parting glass, the fox-hunters left us and proceeded home ; the head keeper kindly telling me I might shoot as many Ptarmigan (although, of course, out of season) as I wanted.

I was not so much surprised at his generosity when I afterwards learned that we were at least half a mile on the ground of his Rannoch friend, to whom I am indebted for my case of Ptarmigan and nest.

It had now become so thick and dark, that I could hardly see a gunshot ahead of me.

In order to give the female a chance to return to

her nest, I left the spot, and went in search of a male. Though several were croaking in all directions, it was some time before I could get a shot. At last a chance presented itself at a bird flying past; but as he went on out of sight, though evidently hard hit, I was just looking out for another, when I heard Donald, the keeper, who was some distance behind, shout out that he had the bird; and on going up to him I found it had fallen dead within a few feet of the spot where he stood.

Then, cautiously approaching the last nest, we discovered that the female had returned. She sat very close, and it was not till the keeper put his hand under her, and lifted her up, that she could be induced to fly.

We next took the nest and eggs, and, after securely packing the whole of our specimens, made our way back to where we had left the pony.

On reaching the spot, the gillie pointed out the nest we were in search of after lunch. We had not examined the ground between the legs of the pony, and here the female had sat unmoved till she had been disturbed by one of the pannier-straps falling on her back while the lunch was being repacked.

We had now twelve or fourteen miles of rough travelling to get over before reaching the Lodge; and as the mist was so dense that we could not see above a yard or two before us, I was of opinion that it would be our safest plan to follow a dry gulley down to a burn, which we knew fell into Glenlyon, and, although a few miles out of our road, would be sure to bring us home at last; but as Donald was so exceedingly confident that he could find his way back across the hill,

I at last gave way, though I certainly had great doubts on the subject.

After leaving the rocky ground we made good progress for about an hour and a half, when I noticed that Donald's cheerful countenance began to wear a troubled look, and he at length proposed that we should make casts for a large rock which ought (supposing we were in the right line) to be at no great distance. After searching for half an hour, but without meeting a single mark that would serve to guide us on our way, we found it was no use to proceed any further in the direction we were going, and on retracing our steps we soon got so confused, that even the trusty Donald was forced to confess that he was at a loss to know which way to turn. After wandering about all night we found ourselves, when the mist cleared off at daybreak, within a short distance of Loch Rannoch, being then just a mile or two further from the Lodge than we were when we started for home the previous evening.

A few days later, as I was anxious to know where we had first missed our way, I went over the same ground, when I discovered that after proceeding about a couple of hundred yards, we had in some manner turned round and taken a nearly opposite course.

On passing the nest of the Ptarmigan which had been caught by the terrier, I found she was sitting on only three eggs, the others having been without doubt carried off by the Ravens. While I was collecting the tail feathers, which had most probably drawn their attention to the nest, I heard a harsh croak, and on looking up the Raven was right above my head, only, unfortunately, out of shot.

As the day was clear, I was enabled to thoroughly examine the plumage of the male Ptarmigan with the glasses, and obtained a much finer specimen than the one I had previously shot.

The male and female, together with the nest and eggs, were obtained on the hills between Loch Rannoch and the River Lyon, in the north-west of Perthshire, in the first week of June, 1867.

RED GROUSE.—(Autumn.)

Case 75.

So much has already been written concerning the habits and diseases of this popular game-bird, both by naturalists and sportsmen, by those well acquainted with the bird in its wild state, and also by others whose experience appears to have been decidedly limited, that there is little left to be said on the subject.

One thing alone is certain, viz. that we are as far off as ever from discovering either the cause or a remedy for that disease which seems to make periodical ravages over the moors, attacking the birds with equal severity on ranges where only a few scattered packs are found, as on the most prolific beats.

Grouse would, in my humble opinion, be found to keep their health better and longer if the moors were more evenly shot over.

In some parts, and frequently on the best-stocked beats, there are only a few weeks' shooting in the beginning of the season, when the ground could well stand two or three guns shooting judiciously over it from the 12th of August till the 10th of December.

It has frequently been put forward that the killing down of vermin destroys the balance of nature, and is prejudicial to the well-being of game.

I do not deny that this may be the case where a too heavy stock of game is kept up; but on ordinary moors, where the ground is properly shot over, the vermin must be kept down, or that very balance of nature, which so delights the theorists to talk about, would soon be lost.

Some years back I hired a moor in Perthshire, where the vermin had been allowed to multiply unchecked, and all precautions for the welfare of the Grouse had been neglected.

The first season I rented the ground the four best beats did not yield an average of above fifteen brace the first day that they were each shot over; after this, the average fell to about seven brace a day. This was for two guns. After three years' trapping and carefully looking after the ground, one gun was able to average forty-five brace of Grouse a day for the first ten days' shooting, without counting two or three hundred head of other game.

In conjunction with the vermin-trapping, I consider that the improvement was mainly due to my making a point of observing the two following rules :—

(1) Always to kill down the single old cock Grouse when and where I could.*

---

* I myself treated them as vermin, and shot them for two or three months after the close of the season. This I am afraid some people might consider highly improper.

(2) Always to be on good terms with the farmers and (more particularly) with the shepherds.

When it is considered that a shepherd is over the ground nearly every day in the year,* and, if so inclined, can report anything going wrong,† the advantage of making an ally of him will, I think, be easily recognised.

There is no vermin ever so destructive on a moor as a badly-fed colly dog.

Old birds are frequently caught on their nests, and the young or eggs are bolted whole.

Anyone taking the trouble to look at the droppings of a colly can easily see whether he has been living on eggs or young birds.

If the shepherd has an interest in the game, he will look well after his dogs, and keep them to heel when not working them.

I found it a good plan to give each shepherd one penny a head for every Grouse bagged on his beat. He was then certain to do his utmost to preserve the game and promote the sport.

I have often heard shepherds in different parts of the Highlands complain of the way they had been treated by the shooting tenants ; and when this is the case, it is no wonder that the sport is not so good as it ought to be.

The specimens in the case were shot on the Innerwick Moors, in Glenlyon, in Perthshire, in October, 1865.

---

* Which, with the numerous duties he has to attend to, can certainly not be expected of even the most energetic of game-keepers.

† Which he certainly will, if he knows it is to his own advantage.

GREYLAG GOOSE.

Case 76.

The Greylag Goose is the only representative of its family that remains with us as a resident throughout the year. Great numbers of these birds rear their young in the more remote parts of Ross-shire, Sutherland, and *C*aithness, and also on some of the surrounding islands.

Though proverbially one of the wildest of fowl during the winter, those that nest on our shores lay aside their shyness while their young are unable to provide for themselves.

While crossing the moors in the summer, I have now and then seen an old gander leave the cover which grew near some small loch, and with outstretched neck attack any dogs I happened to have with me; the female at the same time being either heard or seen endeavouring to get the young into some place of safety.

When on the wing, even at a distance, this Goose may readily be distinguished by the conspicuous grey feathers from which it takes its name.

In 1862 and 1863, when living in East Lothian, we were visited every winter by large flocks of Geese, which were, on some farms, so destructive to the young corn, that herds were obliged to be employed to keep them from the crops.

They take little notice of the labourers while at work; but, though appearing to be all intent on searching for food, the slightest sign of danger is almost certain to attract the attention of the sentinel on duty.

The specimens in the case were obtained in the west of Ross-shire, in May, 1868.

PTARMIGAN.—(Autumn.)
Case 77.

By this time the male has changed the showy dress he wore in the courting season, and, clad in sober grey, matching in colour the rocks and stones he frequents, he gives timely warning of approaching danger to his unsuspicious brood.

On fine days at the commencement of the season it is occasionally no easy matter to force the young birds to take wing, as, unaccustomed to the sight of anyone more terrible than a wandering shepherd, they run like chickens a few yards in advance of the sportsmen.

In wet and windy weather their nature seems entirely changed, as, unless surprised in some rocky corry, no bird is more difficult to approach within gunshot.

The specimens in the case were shot on Benderich, one of the hills to the north of Glenlyon, in Perthshire, in the months of August and September, 1865.

CORMORANT.—(Immature.)
Case 78.

As Cormorants in the immature dress may be seen at all times of the year, it is certain that they must be at least two, and possibly three, years old before they assume the adult plumage.

Though most persons would fancy this bird unlikely

to make a savory dish, they are, in the immature state, by no means unpalatable when properly cooked.

One spring when I was stopping at Canty Bay, I went into the kitchen at the inn, where Adams, the landlord —who then rented the Bass Rock — was getting his dinner. At his request I was helped to (what he called) a plate of Hare soup. There was no denying the fact that I have seldom tasted better soup, but I hardly believe that I should have fancied it much at the time if I had known it had been prepared from certain portions of two *Cormorants* and a Shag that I had shot near the Bass a few days before.

Though using the Bass Rock as a roosting station in great numbers during the winter and early spring, none of these birds have ever been known to breed on the rock.

The specimens in the case were shot at the Bass Rock, in September, 1874.

SHAG.

Case 79.

This wild-looking bird is common round many parts of our islands, usually being found more numerous where the coast is steep and rocky.

Single pairs of this species are occasionally found breeding by themselves, as in the caves at the Bass Rock, and on the " Pinnacles " at the Fern Islands, but more commonly they nest in colonies of smaller or larger size. They used formerly to be plentiful at the Ferns, but during the last two years that I have visited the Islands, there was but one pair.

The specimens in the case, together with their nest and eggs, were taken from the rocks on the west coast of Ross-shire, in May, 1868.

PTARMIGAN.—(Winter.)

Case 80.

It is only the oldest birds that assume the pure white dress so early as the end of the shooting season, the young occasionally retaining several grey feathers in their plumage a month or even six weeks later; and, judging from my own experience, I should be of opinion that some of the more backward birds do not become thoroughly white till their second winter.

This change is not a moult; the white appears first at the point of the feathers, and then gradually spreads down to the root or quill.

I have now and then killed Ptarmigan which at first glance appeared perfectly white, but on being more closely examined showed several stains or lightly-marked blotches on their plumage, and on turning back the feathers I have discovered that a few were still half grey, and their darker colour showing dimly through the pure white covering of the adjoining or overlapping feathers, gave the stained appearance to their otherwise spotless plumage.

Ptarmigan shooting in December is a very different sport to the slaughter of the innocents in August. The hills have now put on their winter covering of snow and ice, and a good bag of White Grouse is seldom made without considerable risk. This is particularly the case in the north-west of Perthshire, where the hills are high and the beats extensive. It is a matter of small importance to be lost all night on a Ptarmigan hill during the summer months, but to miss one's way on a winter's evening, with a snow-storm coming on, might possibly be attended with serious consequences.

In order to avoid such a mishap I used, when making an expedition for these birds, to leave the Lodge by 3 or 4 o'clock in the morning, and so reaching the high ground about 8 o'clock (which was as early as it was possible to commence shooting) I was enabled to get four or five hours' sport, and then have the advantage of daylight for the roughest part of the homeward journey.

Frequently after scrambling up-hill in the dark, we have discovered on arriving at the spot where we expected to meet with the birds, that the whole of the summits were enveloped in cloud and mist. Under such circumstances, all sport of course being out of the question, the only thing left to be done would be to beat a speedy retreat, and hope for more propitious weather on the next attempt.

The specimens in the case were obtained on the hills to the north-west of the River **Lyon**, in Perthshire in December, 1867.

At this season the males may readily be distin-guished from the females by their showing a black mark between the beak and eye.

BEAN GOOSE.
Case 81.

Although some authors have stated that this Goose nests in the British Islands, I believe that no properly authenticated eggs have ever been obtained.

In the North they appear in large flocks early in the autumn, and on the approach of cold weather gradually make their way south.

In East Lothian and other parts of the south of Scotland these birds, as well as the Greylags, become a perfect nuisance, from the depredations they commit on the crops.

When living in that district I was proceeding one evening to dine with a neighbouring farmer, when I perceived what I imagined, in the gloaming, to be a large flock of sheep advancing over a field of young corn. As I knew they had no business there, I went cautiously round the back of the hedge to learn, if possible, where they were breaking through, and on looking over was almost as much surprised as the birds themselves, to discover my flock of sheep transformed into about 500 Wild Geese.

I was totally unprepared at the moment to fire a gun which I had with me, but before they got out of range I managed to bring down two and wound another, which was captured alive by a sheep-dog the following day.

Had I known the birds were there, they were so closely packed that at least ten or a dozen must have been bagged.

The two killed were both of this species, but the other, which I did not see myself, appeared from the description given me to be a Greylag.

I have seen these birds particularly numerous during some winters in the grass marshes in the east of Norfolk. In this locality, the ditches being sufficiently wide to be navigable by a small punt, first-rate sport may occasionally be obtained at them.

When the weather remains open and the supply of food is plentiful, they attain a great weight; a dozen I bagged one day in Norfolk averaging over 9lbs. each, the heaviest just turning the scale at 9¾lbs.

The specimens in the case were shot in the Hickling. Marshes, in the east of Norfolk, in January, 1872.

WHITE-FRONTED GOOSE.

Case 82.

Like the preceding species, this handsome bird is only a winter visitor to the British Islands.

Twenty years ago I have seen hundreds in a day in Pevensey Level, but at the present time a winter often passes without a single bird being obtained in that quarter.

It associates with the " Bean " and other species of Wild Geese.

The specimen in the case was shot on the Holmes Marshes, in the east of Norfolk, in January, 1872.

FRENCH PARTRIDGE.

Case 83.

This fine, handsome bird is by no means a favourite with the generality of sportsmen.

Its well-known shyness and constant habit of trusting to its legs rather than its wings as a means of escape from danger, causing its rapid increase in some counties to be regarded as anything but acceptable.

In the early spring, numbers are frequently picked up drowned in the broads in Norfolk and Suffolk : this is regarded by some of the natives as a proof that fresh arrivals take place at that season ; but I myself, having often noticed their quarrelsome disposition, believe that while flying in pursuit of one another over the water, they become confused, and falling, are unable to regain the shore.

Part of the specimens in the case were picked up drowned on Heigham Sounds, in the east of Norfolk, in March, 1873, the remainder being killed in the Heigham Marshes the following December.

RINGED GUILLEMOT.

Case 84.

It has usually been, I believe, an open question among scientific naturalists, whether this bird is simply a variety of the Common Guillemot or a true species.

Without attempting to give an opinion either way, I consider it strange, if it is a variety, that specimens are never obtained with a partial ring or bridle ; not a single instance having ever occurred where a bird so marked has been obtained.

It has been stated, I believe, on good authority, that this bird has been observed paired with the common species.

I have obtained specimens in the English Channel, both in spring and winter, and they may be found in small numbers at all the nesting stations of the other rock birds, breeding on the same ledges, and mixing with the *Common Guillemots.*

The specimens in the case were obtained partly at the Bass Rock, in the Firth of Forth, in May, 1867, and the remainder at the Fern Islands, off the coast of Northumberland, the following month.

COMMON GUILLEMOT.—(Summer.)

Case 85.

The unfortunate habit of believing everyone to be honest till proved to be otherwise, and trusting them accordingly, has gained for this confiding bird the name of "Foolish Guillemot."

This, if not the commonest, is one of the most numerous of our sea birds, breeding in large colonies at hundreds of stations round the British Islands.

Though not usually considered fit for table, they are extensively used as an article of food on some of the barren islands in the North.

The parent birds are stated (though I have never had the good fortune to be myself a witness of the proceedings) to carry down their young on their backs from the ledges on which they are hatched. Numbers that are knocked over by accident fall into the water, but appear to receive no injury, unless striking against the rocks in their descent.

I reared one myself that dropped over two hundred feet from one of the highest ledges on the west side of the Bass Rock, and only missed the gunwale of my boat by about half a yard.

Though the eggs of the Guillemot differ in a most

remarkable degree, the ground colour being occasionally blue, green, yellow, or white, I believe that one bird always lays the same coloured eggs.

I once removed three eggs from a small ledge on the Bass Rock, and visiting the same spot, about ten days later, I found three more in precisely the same situation, and so exactly like the former ones as to be hardly distinguished from them. Again returning in a fortnight, three more, similar in colour and corresponding almost mark for mark with the others, were obtained. No other Guillemots were breeding within twenty or thirty yards of that part of the rock, and though I frequently examined the spot from the sea through the glasses, I never noticed more than the three pairs frequenting that ledge.

The specimens in the case were all obtained at the Bass Rock, in the Firth of Forth, in June, 1867.

PARTRIDGE.

Case 86.

It is a curious fact that Partridges and Pheasants, if driven over water or towns, appear to get bewildered, and, losing all power of flight, drop down, and suffer themselves to be picked up, rather than rise again.

Some people have a mistaken idea that a land bird is unable to rise from water, but I have repeatedly seen several species that have fallen wounded rise from either fresh or salt water with the greatest ease.

The first year that the Easter Volunteer Review was held on the Downs in the neighbourhood of Brighton, the wind was from the north, and during the sham fight great numbers of Partridges were disturbed by the crowds and noise, and, becoming confused, flew out to sea, where they fell into the water. Several boats which happened to be under the cliffs profited by their misfortunes, one alone getting between twenty and thirty birds. Next year over a score of boats were on the spot awaiting the coming of the unfortunate Partridges, but the wind luckily blowing from the south, carried the affrighted birds inland, and not one came out to sea.

I was going early one morning to the station at St. Leonard's, when I observed a covy of ten or a dozen Partridges drop in a small open square in the back part of the town. On being chased by some boys and dogs, they never attempted to use their wings, but sought shelter in the open doors or fluttered helplessly down the areas.

The specimens in the case were shot at Potter Heigham, near Yarmouth, in Norfolk, in December, 1873.

BRENT GOOSE.

Case 87.

Immense flocks of Brent Geese make their appearance about the end of September or beginning of October on the mudbanks in the firths on the north-east coast of Scotland. In the Dornoch and Cromarty Firths they perhaps collect in the greatest numbers.

On their first arrival they are remarkably un-suspicious of danger, but, from being constantly harassed by the punt gunners, they soon gain wisdom and learn to provide for their own safety. Some, seasons back, over 1,800 were bagged in six weeks by one gunner alone in the neighbourhood of Invergordon in Ross-shire.

In severe weather, they frequent all the muddy harbours and creeks on the eastern and southern coast, and here they also meet with another warm reception, thousands falling victims to the swivel guns, frequently twenty or thirty, and occasionally as many as fifty or sixty, being obtained at one discharge.

On the approach of spring they again work their way back, and finally take their departure for the summer about the end of March or beginning of April. They seem to prefer the eastern to the western side of the island.

I have never met with these birds on inland waters, except on two or three occasions during heavy snow-storms or protracted rough and stormy weather.

Though feeding almost exclusively on the weeds on the salt water mudbanks, their flesh never acquires a fishy taste, and they are considered to be the finest flavoured of all the Goose tribe that visit our islands.

The specimens in the case were shot at the Little Ferry, a muddy harbour a couple of miles south-west of Golspie, in Sutherland, in March, 1869.

EIDER.—(Mature and Young.)

Case 88.

This fine bird is to be met with round several parts of the Scotch coast, and also in a semi-domesticated state on the Fern Islands, off the coast of Northumberland.

The females that nest here appear to select the neighbourhood of the storehouses and other buildings as a kind of protection from the attacks of the larger Gulls which are always on the look-out for any exposed eggs. I have frequently observed the egg-gatherer (who has charge of all the birds on the islands) stroke them on the back when sitting, and even lift them from their nests without their showing the slightest signs of alarm.

The males sat quietly on the water at a distance of 60 or 70 yards, but did not seem inclined to allow a nearer approach.

The following description of the Eider, which I found in an old History of Scotland, may possibly be interesting if not instructive :—

" In this island (Lewis) there is a rare species of bird, unknown to other regions, which is called Colcha, little inferior in size to a Goose, all covered with down, and when it hatches it casts its feathers, leaving the whole body naked, after which they betake themselves to the sea, and are never seen again till the next spring. What is also singular in them, their feathers have no quill ; but a fine light down without any hard point, and soft as wool, covers the whole body. It has a tuft on its head, resembling that of a Peacock, and a train larger than that of a house cock. The hen has not such ornament and beauty."

The males were shot on the Island of Fidra, in the Firth of Forth, in May, 1867, the female and brood being obtained near the Island of Ebris about a couple of miles further west during the following month.

PHEASANT.

Case 89.

This, the most gaudily-attired of all our game birds, was not originally a native of the British Islands, being introduced into this country between two and three hundred years ago; those turned down in the first instance having been brought, it is stated, from the banks of the river Phasis, in Asia Minor.

The specimens in the case were obtained at Potter Heigham, near Yarmouth, in Norfolk, in December, 1874.

COMMON GUILLEMOT.—(Winter.)

Case 90.

During winter immense numbers of these birds frequent the English Channel, following the large shoals of sprats and other small fish that make their appearance at that season.

A few are occasionally met with soon after the new year in full summer plumage, but the majority retain the winter dress till well on in March, and young backward birds of the previous year even longer.

The specimens were shot at sea, a few miles off Brighton, in December, 1870.

BLACK GUILLEMOT.

Case 91.

Though a comparatively rare bird off the English coast, the Black Guillemot is abundant all round the north of Scotland and the adjoining islands.

A few breed, or rather did, some years ago, at Flamborough Head, and it is stated to have been common at one time at the " Ferns." I obtained a single specimen off those islands in May, 1867, but none had then been known to nest there for many years.

This bird differs in several respects from the rest of its family. The Common, or Foolish, Guillemot is a remarkably poor walker, the backward position of its legs only enabling it to progress with a kind of shuffle. Its limited powers of locomotion are, however, sufficient for its requirements on land, as the situation in which it rears its young, viz. an open ledge above the sea, allows it to drop from the air close to its egg. The Black Guillemot, on the other hand, nesting, as it does, either under large stones, or at some distance in the cracks and crevices of the rocks, has need to make use of its legs with greater freedom. I was much surprised the first time I met with this species to discover that it could walk and even run with the greatest ease. It is also capable of rising from level ground (as I observed on seeing a pair disturbed from under some large detached rocks on the Island of Fura) with almost the rapidity of a Grouse or Partridge, while the unfortunate Willock* is forced to drop a considerable distance from the cliffs every time it gets on wing, before it can gain sufficient impetus to take a straight course. This species usually lays two, the other Guillemots a single egg.

The specimens, together with their eggs, were obtained on the Island of Fura, off the west coast of Ross-shire, in May, 1868.

---

* Sussex name for the Common Guillemot.

EIDER.—(Nest.)

Case 92.

This case, which represents the female with her nest, showing the eggs embedded in the well-known Eider down, is copied from a sketch taken on the Island of Fidra, in the Firth of Forth. The nest was placed on a small ledge in the face of the rock, at the height of forty or fifty feet above high-water mark.

I have never visited the island during the spring or summer without discovering a nest in that identical spot; on the last occasion, in 1874, there were two placed side by side.

The female was shot in Gullane Bay, in the Firth of Forth, on May, 1867, and the nest and eggs were obtained on the Island of Fidra the following month.

EIDER.—(Autumn.)

Case 93.

I think it probable that the male Eider is at least two, and possibly three, years old before he assumes the full, white backed plumage.

During summer, young males may be seen in various stages of plumage, the white becoming more conspicuous by age.

The case shows a male killed in September, and a pair of young birds in their nestling feathers.

The specimens were shot in the Firth of Forth, near the Island of Ebris, in September, 1874.

GARGANEY.

Case 94.

This beautiful little Duck is only a summer visitor to Great Britain.

It is far from uncommon in Norfolk and Suffolk, arriving about the latter end of March or beginning of April. Several pairs usually nest in the reed beds and rough marshes in the neighbourhood of Hickling Broad

The young in their first plumage are by no means unlike the Common Teal of the same age, but the pale blue feathers on the wing at once indicate their species.

The specimens in the case were shot on Hickling Broad, in Norfolk, in May, 1870.

GREENSHANK.—(SUMMER.)

Case 95.

The Greenshank makes its appearance in the spring, arriving generally early in May, on its way to its breeding quarters in the far north, and visiting us again in the autumn on its return.

Several pairs, however, remain and rear their young in the northern counties of Scotland. This is more frequently the case than is usually supposed in Ross-shire, Sutherland, and Caithness; most writers asserting that its breeding in the British Islands is a rare occurrence.

The nest is placed on the open moor, at some distance from the usual haunts of the bird.

The female sits remarkably close, and on two or three occasions I have lifted her off her eggs.

As soon as the young can fly they join in flocks, and come down to the shores of the firths.

I have found them particularly numerous on the muddy islands at the head of the Cromarty Firth, near Dingwall, in the beginning of August. They are then in first-rate condition, and few birds, even of the Snipe or Plover kind, are finer flavoured.

I have never met with this species during the winter months.

The specimens in the case, together with their nest and eggs, were obtained in Strathmore, in Caithness, in June, 1869.

JACKDAW.

Case 96.

Jack is generally supposed to be a mischievous rogue, but I had always believed that his character, like that of another black party, was not so bad as it was depicted

One summer, however, when living in Perthshire, I required a young bird of this species, in order to compare with the young of the Grey Crow; and on examining some nests I discovered the shells of dozens of Grouse eggs, which had been destroyed.

It was too late that season to do much good by exterminating the colony, as the mischief was already accomplished, but the next spring I took forcible measures to prevent them from breeding in their old quarters, and the following season the Grouse on the adjoining beat were nearly doubled. The whole of the ground, within about a mile of the rocks where the Jackdaws nested, had till now been perfectly worthless, never more than a few pairs of barren birds being found there.

I have never myself detected them doing much damage to game in England, though I have been assured by shepherds and keepers that they occasionally manage to search out the early Partridge nests on the downs in the south.

Numbers of these birds arrive from the north of Europe in the autumn; I have repeatedly met with large flocks during the month of October, when on their passage. Several times after a gale at that period, I have seen these birds floating either dead or dying on the water, not having had strength sufficient to complete their journey.

The specimens in the case were shot at Offham Chalk Pit, near Lewes, in Sussex, in March, 1872.

R A V E N.
Case 97.

The Raven is distributed over the country from Sussex to Caithness, though considerably more numerous in the northern than the southern parts of the Island.

It nests either in high trees, old buildings, cliffs or precipices, accommodating itself to the neighbourhood it inhabits.

In the beginning of September, I have often seen as many as fifty or sixty of these birds gathered together on the moors in Perthshire; few, if any, of these had been bred in the immediate neighbourhood, having in all probability crossed the hills fron the northern counties or the western islands.   At that time of year they were perfectly harmless on the ground, preying only on wounded game or hares, and, as they always left that quarter before the breeding season, their visits were beneficial rather than otherwise.

On several occasions after a large hare drive, when going over the ground on the following day to pick up the wounded, and also to learn what vermin were about, I have observed them collected in still larger bodies, a hundred or more being scattered over the hills within view, having been drawn from all the sur-rounding moors by the prospect of abundant food.

They were at all times so eager to make a meal off the dead game with which we baited our traps, that I have known between two and three hundred captured in a single season, not that we wished to destroy them, but simply that they positively insisted on getting into the traps which we were forced to keep set, in order to check the increase of more destructive vermin.

In the breeding season there is no doubt that they are injurious to game, being very partial to eggs.

The specimens in the case were trapped in Glenlyon, in Perthshire, in September, 1866.

GREENSHANK.—(Autumn.)

Case 98.

This case represents the mature and immature in autumn plumage.

In the spring the old birds are remarkably wary, but in the autumn they appear to lay aside their shyness, and, when discovered on the mudbanks in company with Redshanks and other Waders, generally fall easy victims to the punt-gunners.

Their loud whistle is often distinguished at night, among the calls of the various species of mudbirds which may be heard during stormy weather in the autumn.

The specimens were shot on Breydon mudflats, near Yarmouth, in September, 1872.

SHOVELLER.—(Mature.)

Case 99.

This handsome bird may be found from north to south, though nowhere particularly abundant : it nests in several different counties, and its numbers receive considerable additions from the continent in the beginning of the winter.

I was particularly unlucky with the finest drake of this species I ever killed. Just at daybreak I made out a pair feeding among the water-plants on Heigham Sounds, and, sculling quietly up within distance, I fired with good effect, stopping both birds, but on proceeding to the spot I discovered that the oakum wad of the punt-gun having struck the male on the neck had completely blown away his head, the remains of which I found lying with the wad about twenty yards distant from the body. The colours in that specimen were by far the brightest and most clearly defined that I ever noticed.

The pair in the case were shot on Hickling Broad, in Norfolk, in December, 1872.

and want of food as to be incapable of flying; numbers at the same time being seen hovering over the breakers a short distance at sea. They kept passing for nearly a fortnight, few being noticed on fine days, but several shewing themselves in the small pools near the sea-beach in rough or windy weather.

They are at all times most unsuspicious of danger, generally allowing themselves to be approached within three or four yards without exhibiting the slightest signs of fear.

The specimens in the case were shot among the Oyster Ponds, in Shoreham Harbour, near Brighton, in October, 1870.

ROOK.

Case 102.

This is one of our most familiar British birds. Opinions differ as to whether the Rook is beneficial, at all seasons, to the farmer and the game-preserver. I am, however, satisfied that the injury it occasionally inflicts on the crops is amply atoned for by the assistance it renders in ridding the ground of worms and grubs. On the other hand, there can be no denying the fact that a nest of eggs in an exposed situation will be as readily destroyed by a Rook, as by that well-known robber the *C*row.

I have, when fishing and shooting in the North Sea during October, often met with large flocks of Rooks on their way to this country. It was seldom that they flew in straggling parties, like the Grey *C*rows; those that were seen singly, appearing to have fallen out from the ranks through fatigue. After a gale of wind from the south-west, I have seen several floating dead on the water, between twenty and thirty miles off the land. I have also received a few wings from the light-ships off the east coast, during the winter months, the birds having fallen disabled on deck after striking the lamps. From never having observed them on their return journey in the spring, or obtained any wings from the light-ships at that period, I am ignorant whether they take up their residence in this country, or again return to the North of Europe, from which country they appear to be making their way when met with in the autumn.

On one occasion I heard that, after being banished from the rookery they inhabited, for their depredations on the Grouse eggs, the poor birds nested out on the open moor, trees being scarce in the neighbourhood.

The specimens in the case were obtained near Brighton, in June, 1872.

CUCKOO.—(Mature.)
Case 103.

The Cuckoo, as most people are aware, is only a summer visitor to Great Britain, its arrival being eagerly looked for as one of the signs of approaching spring.

It is common all over the country, its well-known note being heard as frequently on the wild moors of Sutherland, as on the furze-covered downs of Sussex.

Its curious habit of entrusting its egg to the care of other species is so generally known as to need but a passing mention. A long list of these foster-parents is given by many authors, the commonest being, in my opinion, the Meadow Pipit, Wagtail, and Reed Warbler.

In some districts the natives have an idea that during the winter this bird turns into a Hawk. I have been gravely assured that specimens showing the change have been frequently obtained, only unfortunately not preserved.

The specimens in the case were shot on the Downs, near Brighton, in May, 1870.

CUCKOO.—(IMMATURE.)
Case 104.

In this case the young bird is represented as being fed by its foster-parent, the Titlark or Meadow Pipit.

The specimens were obtained on the hills near Lewes, in Sussex, in June, 1874.

POCHARD.
Case 105.

A few of these birds occasionally nest in the British Islands, but it is decidedly the exception, not the rule.

Immense flocks make their appearance during the autumn on the lochs in Scotland, and on the lakes and broads in England. When not molested they become remarkably fearless, but from being so generally persecuted they soon get wild, and it needs no little skill to work a punt within gun-shot. They may, however, generally be approached with some chance of success just before daybreak; they then seem disinclined to take wing, drawing all together and swimming in a compact body.

I had made a very successful shot at these birds a few winters back in the east of Norfolk, and the following morning I was again on the look-out for the flock. Though prevented by a thick fog from finding them for a considerable time, I at last caught sight of what I took to be nearly a hundred swimming close together, at about sixty yards' distance, and bringing the gun to bear—my finger was on the trigger—when suddenly a head appeared in the centre of the object, and I discovered that the flock of Pochard was a man in a punt, within ten yards of the muzzle of my gun. One second longer and the unfortunate fowler would have received the contents of a punt-gun loaded with a pound and a quarter of shot. I nearly got peppered myself on one occasion under similar circumstances, and, considering the imperfect light in which the heaviest shots are generally made, it is a wonder that more accidents do not occur.

The flesh of this bird is supposed to resemble that of the celebrated American Canvas-back Duck.

The specimens in the case were shot on Loch Slyn, in the east of Ross-shire, in March, 1869.

PAGETS POCHARD.

Case 106.

It is considered that this bird is a cross between the Common Pochard and the White-Eyed Duck. On carefully comparing it with the two species, I think there can be but little doubt on the subject.

It was quite by chance that the present specimen was obtained. I had pulled the trigger of the punt-gun at a bunch of Dunbirds* on Hickling Broad, when the charge hung fire for a second or two, exploding just as the Pagets Pochard happened to be flying past the line of the shot. Had the charge ignited when I pulled the trigger, it must have escaped with a fright alone.

As stated above, this bird was obtained on Hickling Broad, in the east of Norfolk, in November, 1871.

ARCTIC TERN.—(MATURE.)

Case 107.

Like all the rest of its family, the Arctic Tern is only a visitor to the British Islands, arriving in the spring, rearing its young on our shores, and leaving us on the approach of cold weather.

There are still several breeding stations at different points round the coast, though two or three spots, where they formerly nested in the South, have been of late years entirely deserted.

Great numbers still take up their summer residence on the Fern Islands. The sketch from which the case is copied was taken at one of their colonies, close to the old lighthouse in the centre of the group.

I have seen these birds nesting in the oat fields in the neighbourhood of some of the Scotch firths, on the north-east coast of Scotland.

The specimens in the case were shot on the shores of the Dornoch Firth, near Tain, in June, 1869. The eggs were obtained at the Fern Islands, in June, 1867.

---

* The Dunbird, or Sandyhead, is one of the names by which the Common Pochard is known among the gunners on the South Coast. A bunch signifies a small flock.

JAY.

Case 108.

This sprightly bird is fast disappearing from our woods and thickets. An egg being a bait it can never resist, it speedily falls a victim to the watchful gamekeeper.

Many writers lament the scarcity of this and numerous other handsome though destructive birds, forgetting that the increase of their favourites signifies a corresponding diminution of all species of game.

The specimens in the case were shot near Plumpton, in Sussex, in February, 1870.

MAGPIE.

Case 109.

Though becoming scarcer every year, like the Jay, the Magpie is still sufficiently common to be well known in most parts of the British islands.

During winter it occasionally joins in large flocks. I have more than once met with as many as thirty or forty in company in the neighbourhood of Doncaster, in Yorkshire.

It is generally a shy, wary bird, always appearing conscious that its bad deeds will gain it but little favour in the sight of man.

I have, however, in some districts in the north, and again in the west, found it remarkably confiding, even nesting in trees within a few yards of a dwelling.

The specimens in the case were shot on the downs in the neighbourhood of Brighton, in the spring of 1872.

ARCTIC TERN.—(Mature and Immature, Autumn.)

Case 110.

During August and September large numbers of these birds may be met with along the southern and eastern coasts while on their journey to their winter quarters.

The case represents an old bird in the change from the summer plumage, and the young of the year.

The specimens were shot at sea, off Benacre Sluice, a few miles north of Southwold, in Suffolk, in August, 1873.

SCAUP DUCK.

Case 111.

When living in the Highlands, I have occasionally noticed a few pairs of Scaups remaining so late in the summer, that I fancied they must be nesting somewhere in the neighbourhood : the drakes being often observed without the ducks, strengthened the belief; but, from never having had time to thoroughly watch them, I can state nothing with certainty.

They arrive in this country about the same time as the Pochard. Though decidedly more marine in their habits, they nevertheless often make their appearance on inland waters in both England and Scotland.

They are known among the gunners on the south coast as Frosty Back Wigeon.

The specimens in the case were shot in Edderton Bay, in the Dornoch Firth, in March. 1869.

BAR-TAIL GODWIT.—(Spring.)

Case 112.

The Godwits are here shown in the intermediate state between winter and summer plumage.

When they first arrive on the mudbanks in the southern and eastern counties early in May, the greater portion of them are in this stage ; but as the season advances the full-plumaged birds become more numerous.

The specimens in the case were shot on **Breydon** mudflats, near Yarmouth, in the first week in May, 1873.

BLACKTAILED GODWIT.

Case 113.

Though formerly breeding in great numbers in the eastern part of the island, the Black-tailed Godwit is now only an occasional visitor in spring and autumn.

Small parties still frequent the flat district round the broads in Norfolk and adjoining counties in the end of April or beginning of May, but after remaining a few days they take their departure. In August and September they again appear on their way to the south. By far the greater part of those observed at this season are young birds of the year.

During the autumn they were far from uncommon on the marshes and mudbanks near Rye, in Sussex, sixteen or eighteen years ago; but I believe their visits have now entirely ceased.

The specimens in the case were shot on **Breydon** mudflats, near Yarmouth, one in August, 1871, the other in September, 1872.

BLACK CAP.

Case 114.

The Black Cap arrives in the spring, passes the summer in our plantations and gardens, and leaves in the autumn, soon after its young are sufficiently strong to provide for themselves.

It has been stated that a few remain through the winter in the western part of the island; but I have never met with any at that season.

The song of this bird is but little inferior to that of the Nightingale.

The specimens, both old and young, were obtained at Portslade, near Brighton, in June, 1874.

WOODWREN.

Case 115.

The Woodwren is another summer visitor. I have noticed this bird as being particularly numerous in the wildest glens of Perthshire, Ross-shire, and Caithness.

When observed in the south, in the neighbourhood of Brighton, it appears to have a partiality for high trees, especially beech; the fine old timber in Stanmer Park being one of the favourite haunts of this bird. In the north, however, it is forced to put up with the stunted birch and fir that alone appear to thrive in the rocky glens it frequents. The nest is placed on the ground, generally, at no great distance from the roots of a tree.

The specimens were obtained in Glenlyon, in Perthshire, in June, 1867.

CROSSBILL.
Case 116.

The visits of this curious bird to the south of England are very uncertain.   In the northern counties it appears regularly every winter; and in several parts of Scotland it is a permanent resident.

The *C*rossbill breeds early in the season.   On two or three occasions I have seen large flocks of several hundreds frequenting the fir woods of Ross and Sutherland in the beginning of July.

The specimens in the case were obtained near Beauly, Inverness-shire, in July, 1876.

NIGHTINGALE.
Case 117.

Those unaccustomed to a country life are often surprised that this noted songster is not a finer or more conspicuous bird.

The first arrivals usually take place in the beginning of April.   The song is then continued for a month or six weeks, after which it is heard no more, the bird simply giving notice of its presence by its curious croaking note of warning to its young.

Great numbers are annually taken in traps, on their first appearance in the spring, though but a small percentage of these are ever reared.   A few, I suppose, occasionally sing in captivity ; but all that I have ever seen were such a wretched mockery of the bird, in the state in which I have been used to watch it, as to be positively painful to look at.

The specimens in the case were taken near Plumpton, in Sussex, in April, 1866.

GREEN SANDPIPER.

Case 118.

It is only of late years that naturalists have been acquainted with the singular fact that this Sandpiper lays its eggs in the deserted nests of Wood-pigeons, Thrushes, and other birds, in the branches of trees at a considerable elevation from the ground. It is believed that a few pairs annually rear their young in the northern counties of England, though the greater number proceed to the north of Europe on the approach of summer.

I have once or twice met with this species in winter; it is, however, most common in Great Britain during spring and autumn.

The Green Sandpiper is at all times very difficult to observe closely, taking wing on the slightest signs of danger, and generally flying to a considerable distance.

Two of the specimens in the case are old birds shot in the South Marshes, near Yarmouth, in August, 1871; and the third is a young one killed near Hickling Broad, in July, 1873.

BARTAIL GODWIT.—(Summer.)

Case 119.

Years ago the arrival of the "May-birds," as these Godwits were called, used to be anxiously looked for by all the shore gunners round our southern and eastern coasts. On their first appearance they were generally exhausted by their long flight, and fell easy victims to their numerous assailants.

SHOVELLER.—(Immature.)

Case 100.

This case shows a male in the dingy plumage it exhibits after the breeding season, and the young birds in that stage when it is first possible to distinguish the immature males from the females : the white and red feathers on the breasts of the males just shewing through their nestling plumage.

The specimens in the case were shot on Hickling Broad, in Norfolk, in the autumn of 1872.

GREY PHALAROPE.

Case 101.

A few of these birds visit our coast every autumn. After unusually severe gales, in September or October, they are occasionally found in immense numbers. Should the storm continue many days they seem to suffer greatly from its effects, and scores may be seen so exhausted that they allow themselves to be caught by hand.

In September, 1866, a large flight appeared all round the eastern and southern coasts.

I was myself in the Highlands at the time, and obtained a single specimen in Glenlyon, in Perthshire, at a distance of forty miles from the sea.

In the middle of September, 1870, I observed a few passing along the south coast; and after a heavy south-east gale, about three weeks later, hundreds were found in the neighbourhood of Brighton, and the flight extended as far west as Plymouth. I picked up nearly a dozen, one morning, so disabled by the storm

If all the yarns of the old Breydon fowlers are to be believed, these birds must in days gone by have visited the mudflats in that district in countless thousands. At the present day their numbers have sadly fallen off, as I have often spent the whole of May, from daylight till dark, on the mudbanks, without seeing more than two or three hundred pass during the month.

The flight time commences about the beginning of May, the 12th to the 15th being usually considered the best days; the date, however, varies with the wind, east-south-east, east, and east-north-east being the most favourable quarters. Should the wind continue west or south-west during the whole of May, it is quite possible that hardly a bird will be seen. When this is the case, their line of flight appears to miss our shores entirely. I have on two or three occasions observed large flocks passing twenty or thirty miles from land; and some of the fishermen, who were old gunners, have assured me that they have met with all the different species of Waders in continual flights for several days outside the Dogger-bank, while their usual quarters on the mudbanks were completely deserted.

The specimens in the case were shot on Breydon mudflats, on May 12th, 1871.

BAR-TAILED GODWIT.—(Autumn.)

Case 120.

This case represents the young as they appear on their first arrival in this country, from their breeding-grounds in the far north.

Like the greater part of the birds that are reared in those deserted regions, they are of a most confiding disposition, appearing, unfortunately for themselves, to be totally unacquainted with the destructive propensities of man.

The specimens in the case were shot on the Island of Ebris, in the Firth of Forth, in September, 1874.

COMMON SANDPIPER.—(SUMMER.)

Case 121.

This bird is only a summer visitor to the British Islands, arriving in April, and leaving, after rearing its young, in the autumn. The nest is usually placed amongst the stunted vegetation on the banks of some of our northern streams and rivers.

The Sandpiper is able to swim and dive with the greatest facility. While fishing on the river Lyon, in Perthshire, one of these birds, pursued by a Merlin, dashed into the water within a few feet of where I was standing, and immediately dived beneath the surface, probably reappearing under the cover of some over-hanging bushes, as I did not catch sight of it again.

The specimens in the case were obtained on the river Lyon, in Perthshire, in June, 1866.

NIGHTINGALE.—(AUTUMN.)

Case 122.

The young birds are here shown in their nestling plumage.

The specimens were obtained near Hayward's Heath, in Sussex, in July, 1874.

GREAT TITMOUSE.

Case 123.

In England this handsome bird is dispersed over the country from north to south. In Scotland it is but seldom observed in the more northern counties.

The specimens in the case were obtained at Portslade, near Brighton, in June, 1874.

LONG-TAILED TITMOUSE.

Case 124.

The elaborately-constructed nest of this species is well known to even the most juvenile of egg collectors. Its position, however, varies considerably, being occasionally discovered within a yard of the ground in some prickly furze bush or thorn hedge, or at a height of thirty or even forty feet in the branches of some lofty tree.

These birds are seldom noticed singly during autumn or winter, the families of the preceding summer keeping company till the approach of the following spring. They also occasionally join in considerable flocks.

The specimens in the case, together with their nest, were obtained near Plumpton, in Sussex, in April, 1866.

BLUE TITMOUSE.

Case 125.

The Blue Tit is a most courageous little bird, the female generally resenting an approach to her nest to the utmost of her ability, and occasionally attacking the intruder with the greatest ferocity.

The male and female in the case were captured in a butterfly net while attempting to defend their young, who were just on the point of leaving the old stump in which they had been hatched.

The case is copied from a sketch taken from the identical spot in a plantation near Nigg, in the east of Ross-shire, in June, 1869.

COMMON SANDPIPER.—(Autumn.)
Case 126.

The specimens in this case are in the immature plumage, having been obtained just previous to their departure, in August 1874, between Lancing and Shoreham, in Sussex.

YELLOW WAGTAIL.—(Summer.)
Case 127.

I have nowhere met with this bird so plentiful as in Norfolk; the marshes in the neighbourhood of the broads appearing to be admirably adapted in every respect to their requirements.

For a day or two on their first arrival in April they may be seen in great numbers on the coast of Sussex, alighting on the freshly-turned land wherever the ploughs are at work, running nimbly over the clods within a yard or two of the teams, searching for insects and other food.

The specimens in the case were obtained in the Heigham Marshes, near Yarmouth, in June, 1870.

YELLOW WAGTAIL.—(Immature, Autumn.)

Case 128.

Thousands of young Wagtails in this state of plumage may be observed in the marshes on our eastern coast shortly after midsummer.

The specimens in the case were obtained in the Hickling Marshes, in Norfolk, in July, 1873.

GREY-HEADED WAGTAIL.

Case 129.

The Grey-headed arrives about the same time as the Yellow Wagtail, though in considerably smaller numbers.

A few pairs are said to occasionally breed on the British Islands, though I have never myself had the good fortune to meet with a nest.

The strongly marked grey head of the male causes him to be easily distinguished from the common species, though the difference in the female is so slight that it can only be detected on careful examination.

The specimens in the case were obtained between Brighton and Portslade, in April, 1874.

PIED WAGTAIL.—(Winter.)

Case 130.

As will readily be seen from an examination of the specimens in this case, the Pied Wagtail, although retaining the same colours as in summer, is considerably altered in appearance during the winter months.

The birds were obtained in the immediate neighbourhood of Brighton, in December, 1874.

PIED WAGTAIL.—(Summer.)

Case 131.

A few of these birds remain with us during the winter, though by far the greater number are only summer visitors to the British Islands. All through March and the greater part of April they may be observed on fine still mornings landing on the south coast by thousands; they never appear even in the slightest degree fatigued or exhausted by their passage, as, after alighting for a few minutes at some brackish pool, or on newly ploughed land, they invariably continue their journey direct to the quarters they intend to take up for the summer. Although hundreds might have been observed within a mile or two of the shore during the early morning, it is seldom that more than a pair or two will be met with after two o'clock in the day, the whole of the birds of passage having made their way inland.

Should the weather set in cold and stormy, few, if any, will make their appearance, but with a change of temperature, their accustomed haunts will again be alive with fresh arrivals.

The case is copied from a sketch taken on the north side of the Downs, between Falmer and Plumpton, in Sussex, the specimens, both old and young, being obtained at the same place in June, 1872.

GREY WAGTAIL.—(Summer.)

Case 132.

It is but seldom, in the south, that this bird is met with in the summer plumage; its breeding quarters being for the most part in the northern counties, and in the Highlands of Scotland.

When living in Perthshire, I used generally to notice their arrival about the middle of April. Though they usually frequent the stony banks of rivers and rocky burns, I have noticed a few that bred in the neighbourhood, regularly visiting the dust-heaps and open drains that are to be met with in the centre of some of the Highland towns.

The sketch from which the case is copied was taken from the bridge at Innerwick, in Glenlyon, in Perthshire, where a pair of these birds nested annually, in a small hole in the masonry from which a stone had been dislodged.

The specimens, together with their nest and eggs, were obtained at the identical spot, in May, 1867.

Case 133.

WHITE WAGTAIL.

Case 134.

A few of these birds may be observed on the south coast during April and early in May. The greater part of those I have noticed appeared to be making their way from west to east.

There seems to be considerable doubt among naturalists as to whether this is a true species, or only a continental form of our common Pied Wagtail.

The specimens in the case were obtained between Shoreham and Lancing, in Sussex, in April, 1872.

GREY WAGTAIL.—(AUTUMN.)

Case 135.

There are several localities in Sussex and other southern counties where these birds may be met with during autumn and winter.

I have repeatedly observed them in Scotland frequenting the rocks on the sea-shore at this season.

One of the specimens in the case was shot on the banks of the river Ouse, near Lewes in Sussex, in November, 1873, and the other at Canty Bay, in East Lothian, in October, 1874.

KNOT.—(SUMMER.)

Case 136.

The number of Knots that visit our shores during summer appears to have fallen off greatly of late years. This I can in no way account for, as they seem fully as plentiful at other seasons.

Immense flocks formerly showed themselves on the mudbanks on our southern and eastern coast during May, remaining only for a tide or two to rest, and then resuming their journey to their breeding-grounds in the unexplored regions of the far north.

At the present time, however, a few straggling parties of at most a dozen or twenty birds are all that are usually observed; and in the spring of 1871 but half a dozen full-plumaged Knots were seen on Breydon mudflats between the 5th of May and the 6th of June.

The specimens in the case were obtained in Brey don, near Yarmouth—two in May, 1871, and the remainder in May, 1873.

KNOT.—(Immature Autumn.)
Case 137.

The young Knots that visit us in the autumn generally make their first appearance in this country about the end of July, and continue arriving for six or eight weeks longer. In 1868, however, I met with several large flocks composed entirely of immature birds on both the Sutherland and Ross-shire shores of the Dornoch Firth as early as the 8th of July.

Though the Knot is at all seasons one of the most accessible of our mud-birds, the young on their first arrival in the autumn occasionally suffer themselves to be shot at time after time without making the slightest attempt to escape, the survivors of the flock simply rising on wing at each discharge, and, after a short flight, settling again with the dead and wounded.

When shooting on the mudbanks, I always make

use of a number of " dummies."* These are arranged on the most frequented mudbanks; and it is seldom that a flock of waders will pass over the flats without flying round, if not immediately settling to them. On several occasions I have had as many as from three to five hundred birds of various species gathered round my " dummies" within sixty yards of the gunning punt, the nearest probably being within a boat's length.

The arrival of a large flock of Knots is a most amusing sight. After wheeling two or three times round, they invariably alight close by, and, having thoroughly examined the decoys, a general conversation appears to take place; then one by one they thrust their beaks under their feathers, and in a few minutes the whole are resting quietly on one leg. They take but little notice of our moving about in the boats at forty or fifty yards' distance; but, should our actions appear too threatening, a few will probably run up, and attempt to warn the unconscious " dummies."

On approaching still closer, they will occasionally rise on wing; but, finding their wooden friends do not accompany them, they will settle again, and endeavour with loud twitterings to induce them to start.

The Knot is the only wader that I have ever remarked so careful for the welfare of their lifeless companions; but Pochards commonly exhibit precisely

---

Imitation birds carved out of wood, and painted to represent Plovers, Godwits, or Knots; being strongly constructed, and furnished with iron legs and beaks, so as not to be injured if struck by the shot.

the same regard for the wooden decoy ducks which I generally use during the winter months. These are anchored by means of a line and stone, and always bobbing about, head to wind, are such a correct representation of a flock of wild fowl, that a gunner who is a stranger to them is certain to approach and fire; and many a charge have my poor decoys received.

On one occasion, on Breydon mudflats, three men in one boat cautiously worked up to what they considered a fair distance (about twenty yards), and, all aiming together with the greatest care, fired a volley of six barrels into the thick of the devoted " dummies." Two of them had already loaded and fired again, when a large black hairy quadruped—called by his master, I suppose, a retriever—dashed from the boat, and charged into the centre of the supposed game. Then perceiving that the sagacious animal did not commence to shake and worry the birds (which operation I had previously watched him successfully perform on a wounded Dunlin, to the evident delight of his owner), the sportsmen appeared to comprehend the state of affairs, and beat a speedy retreat.

The Knot seems a particularly sociable bird, joining in company with nearly every other species of wader. I remember one autumn on Breydon, that besides above thirty Knots, I obtained at one discharge of the punt gun, specimens, in larger or smaller numbers, of the following species: Ruff, Redshank, Pigmy curlew, Dunlin, Stint, Greenshank, Spotted Redshank, and Golden Plover. They also commonly associate with the Godwit, Grey Plover, and Turnstone.

The specimens in the case were shot on Breydon, in August, 1872.

REDSTART.

Case 138.

About the second week in April numbers of these handsome little birds may be observed flitting along the Sussex hedges : having passed the winter in a warmer climate they are now on their way to their breeding quarters.

A few pairs nest in the wild forest district about Balcombe and Tilgate ; but the species is far from common in Sussex during the summer months. I have noticed them particularly abundant in the wooded glens in the Highlands, where the old stone dikes and rugged weather-beaten trees afford ample choice for the selection of a nursery.

As the autumn advances, they again work their way south, and take leave of us for the winter.

The male was shot at Catsfield, in the east of Sussex, in May, 1855 ; the female in the neighbourhood of Brighton, in April, 1867 ; and the nest and eggs were taken a month later in Glenlyon, in Perthshire.

BLACK REDSTART.

Case 139.

Unlike the common, the Black Redstart is a winter visitor to the British Islands, usually appearing about the middle of October, and taking its departure in March, or the beginning of April.

It is by no means so abundant as the common species, though in a few localities it may generally be observed by those who are aquainted with its habits and manner of flight.

It seems to have a partiality for the rocks and cliffs on the sea coast, being generally found every autumn between Brighton and Newhaven, and again in the neighbourhood of Plymouth.

The mature males are by no means so plentiful as the females and young.

One of the males in the case exhibits the white on the wing, while the other shows not the slightest trace of it, most probably being a younger bird.

The specimens were all obtained along the chalk cliffs between Brighton and Rottingdean, in October, 1873.

GREY REDSTART.

Case 140.

Most authors declare that the Grey Redstart of the British Islands is only the immature of the Black.

I do not pretend to risk an opinion on the subject, and will only state that two of the specimens in the case were stated, by the taxidermist who set them up, to be males.

I have observed numbers of this description of bird frequenting the dust-heaps and newly-spread manure in the neighbourhood of Brighton, from the end of October till the middle of November.

One of the specimens in the case was shot while flying off the roof of the Museum, in November, 1874; and the remainder in the adjoining grounds, in November, 1875: at which time they were particularly numerous.

WHEATEAR.— (SPRING.)

Case 141.

This is one of the earliest of our spring visitors, appearing frequently before the cold weather has left us. It rapidly spreads itself over the country, and during the breeding season is quite as numerous on some of the northern moors as on the Downs of Sussex.

The specimens in the case were obtained early in April, 1866, in the neighbourhood of Brighton.

KNOT.—(WINTER.)

Case 142.

During severe weather Knots are often observed on the mudbanks in flocks of several thousands. I have seen them particularly numerous in the harbours on the coast of Sussex and Hampshire, also in Norfolk, and as far north as the flat shores of Ross-shire and Sutherland.

In consequence of their remarkable tameness, they are certain to fall victims to the punt-gunners, and from one to two hundred are often bagged at one discharge; as they generally bring in to the fowlers from fourpence to fivepence each, a large flock of Knots is always an acceptable sight.

The birds in the case were killed at the Little Ferry, in Sutherland, in March, 1869.*

---

* Seventy Knots and twenty-four Godwits, the whole being the result of a single shot, were picked up together with the present specimens; at least half as many more being swept away by the flowing tide before those nearest at hand could be gathered up. Considering the fact that the gun was only loaded with 10 oz. of No. 1, it was by no means a bad shot.

GREY PLOVER.—(Spring.)

Case 143.

The specimens in the case shew the intermediate stage of the Grey Plover, between winter and summer plumage.

They were obtained on Breydon mudflats early in May, 1873.

GREY PLOVER.—(Summer.)

Case 144.

At this season the Grey Plover is one of our handsomest birds.

Large flocks are occasionally seen on our eastern coast during May, though the numbers that make their appearance are very uncertain.

Their visits are only of short duration. After resting for an hour or two (should the wind be favourable) they mount high in the air and pursue their course, which is almost invariably north-east.

The specimens in the case were obtained on Breydon mudflats, in May, 1871.

These Plovers were particularly numerous that season, successive flocks following one another from the 12th to the 26th of May, after which but a few in imperfect plumage were seen.

WHEATEAR.—(Autumn.)

Case 145.

This is the plumage in which the Wheatear is seen just previous to its departure.

For a month or six weeks in the autumn, during the migration, they are very numerous along the range of the South Downs, and on the marshes adjoining the coast.

The shepherds in the neighbourhood of Brighton used, in days gone by, to catch thousands of these birds in horse-hair nooses on the sheep walks; at the present time, however, there appear to be but few traps ever set, the nets of the bird-catchers taking sufficient to supply the market.

The specimens were obtained near Brighton, in August, 1874.

WHEATEAR.—(Nestlings.)

Case 146.

The young birds are here shown, shortly after leaving the nest, in their first feathers.

They were shot between Shoreham and Beeding, in Sussex, in July, 1874.

WHEATEAR.—(Large Variety.)

Case 147.

A large variety of the Wheatear makes its appearance every season, about three weeks or a month later than the first arrivals of the smaller or common kind.

The difference in size is very conspicuous, and in addition to this the larger when disturbed almost invariably, if possible, flies up and settles on a tree or hedge; while the smaller, under similar circumstances, generally contents itself with a large stone or clod.

Though both breed commonly in this country, I have never noticed them to pair.

The specimens in the case were obtained in the neighbourhood of Brighton, in April, 1870.

BRAMBLING.—(Spring.)
Casc 148.

The present species is only a winter visitor to the British Islands; a few, however, remain as late as April, and often assume the full summer plumage before they take their departure.

In the summer of 1866, while fishing on the River Lyon, in Perthshire, I had occasion to climb a beech tree to release the line which had become entangled in the branches, and while so engaged a female Brambling was disturbed from her nest, containing three eggs, which was placed close to the stem of the tree. As I was anxious to procure the young I left her, and on again visiting the spot in about a fortnight the nest was empty, and, judging by its appearance, I should be of opinion that the young birds had been dragged out by a cat.

This is the only instance I have ever known of the Brambling attempting to rear its young in Great Britain.

The specimens in the case were shot near Falmer, in Sussex, in April, 1875.

KITE.—(Immature.)
Casc 149.

'he specimens, which were shot in Perthshire, in 878, shew the plumage of the young birds as soon as they are able to leave the nest, and before the tail has acquired its full length.

They are here represented as preying on a Grouse, which is, in my experience, their favourite food.

I once counted the remains of over 30 Grouse under the branches of one large Scotch Fir, which stood within a short distance of a nest; some were merely bleached and weather-beaten skeletons, and may possibly have laid since the previous season.

BRAMBLING.—(Winter.)

Case 150.

Immense flocks of these birds make their appearance every autumn in the Highlands, and on the approach of winter gradually work their way south. They often join in company with other small birds, such as *Chaffinches* and Yellowhammers.

The specimens in the case were obtained near Shoreham, in January, 1871.

GREY PLOVER.—(Autumn.)

Case 151.

This case shows the immature birds on their first arrival in this country in the autumn, and also the winter plumage of the adult.

The specimens were obtained on Breydon mudflats, in the autumn of 1872.

MUTE SWAN.

Case 152.

The Mute or Tame Swan being included in all the lists of British birds, is here shewn with its young brood.

The specimens in the case were obtained on Somerton Broad, in Norfolk, in June, 1871. The weight of the male was 32lbs., that of the female 18lbs.

GANNET.—(Mature and Nestlings.)

Case 153.

There are about half-a-dozen breeding stations of these birds round the British Islands. It is, however, only during spring and summer that they approach the land (unless weakened by stress of weather or accident) for the purpose of rearing their young; at other seasons they follow the shoals of herring and mackerel, or other fish on which they feed, and rest on the open sea.

The present case is copied from a sketch taken on the north side of the Bass Rock. The mature birds are here represented, some with their newly-hatched and others with full-fledged young. On breaking the shell the young Gannet is a small, naked, shapeless monstrosity; its first covering is a thin white down, which gradually thickens with the size of the bird; this it retains for about a month before the slightest signs of feathers make their appearance. The nestling plumage, as will be readily seen by the specimens in the case, is a dark grey speckled with white; the back of the head and neck occasionally retaining the down for some time after the other parts. While in their infancy they are the most peevish little wretches, snapping, quarrelling, and fighting with the utmost ferocity. Though the personal injury that they inflict on one another is generally small, their battles are not unfrequently attended with fatal results, as one or perhaps both of the combatants lose their balance, and, falling from their ledge, are dashed to pieces on the rocks below. It is a curious fact that not the slightest notice is ever taken of a young one that drops on the water, even by its own parents, while an old bird that is shot will immediately draw hundreds around it, where they will remain flying in circles till the bird has drifted a mile or two from the rock.

The landlord of the inn at Canty Bay, who also hires the Bass Rock, depends mainly on the geese for paying his rent. The average take for each season is now about eight or nine hundred full-fledged young birds ; this, however, depends greatly on the weather, as, should the rocks be wet and slippery from continued rain, it renders the work of going over to collect them both dangerous and unpleasant, as the liquid guano lies in pools a foot or two deep on some of the ledges and in the cracks of the rocks.

Ten or fifteen years ago as many as 1,500 or 2,000 were occasionally taken, but since that time the birds have greatly decreased.

After being plucked and cleaned, some are sent to the markets at Birmingham, Manchester, and other large provincial towns, and generally bring in from eightpence to tenpence each ; others are hawked about the country, and sold for what they will fetch, while a hundred or two are cooked at *C*anty Bay, and eagerly bought up by the farm labourers of the district for a shilling each.

The fat that comes out of the inside of the birds, when cleaned, is boiled down into oil, and sold for from three to four shillings the gallon ; and, in addition to this, the feathers realise about fifteen or sixteen shillings a stone. These are used for making beds, but have to go through some powerful baking process in order to remove the smell of the guano which clings to them. A roast goose* appears to be a favourite dish with some of the visitors at North Berwick, but the stink of the oil and the cooking at *C*anty Bay set me for ever against such a greasy delicacy.

The specimens in the case were obtained at the Bass Rock, in the Firth of Forth, during the summer of 1874.

GANNET.—(Immature.)

Case 154.

The present case shows the Gannets in the intermediate stages between the nestling and the mature bird. Though the men in charge of the Bass Rock always declare that the Geese are five years old before they assume the full plumage, I think there can be but little doubt that it is attained in the third year.

From having myself kept these birds in captivity for the last two years, I can already perceive that the stage that they describe as two years old is, in reality,

---

* In this district a Gannet is always termed a Goose.

only an early-hatched bird of the preceding season; in like manner I expect that their third and fourth year birds will be discovered to be of the same age.

The graceful manner in which the Gannet takes its prey, steadying itself for a moment in the air, and then darting headlong beneath the waves, must be seen to be thoroughly understood.

The amount of fish that these birds consume is something enormous. It has been estimated that the Bass Gannets number about 50,000, and as each bird will readily devour ten herrings in a day, it can be easily calculated that the market value of the fish required for one day's consumption by this colony alone would be considerably over £1,000.

The specimens in the case, representing the birds at one and two years old, were obtained at the Bass Rock during August, 1874.

BEWICK SWAN.

Case 155.

The visits of this small Swan to the British Islands are very uncertain; hundreds may be observed one winter, and not a single bird make its appearance the following season. They do not generally arrive in such large bodies as the Hoopers; but I once counted between fifty and sixty flying in company over Hickling Broad, in Norfolk.

The specimen in the case was shot on Hickling Broad, in March, 1871; it was a female, and only weighed 9 lbs.

SPOTTED REDSHANK.

Case 156.

The Spotted Redshank, in the black dress of summer, is a rare bird in the British Islands. When shooting on the Norfolk mudbanks, I have occasionally seen a bird or two in May, but at too great a distance to distinguish them, had my attention not been first attracted by their well-known note.

During autumn the immature birds are by no means scarce in some of the eastern and southern counties. Though several make their appearance on Breydon mudflats, I have always noticed them to be remarkably unsociable with their own species, generally being observed singly, or in company with Greenshanks or Common Redshanks.

The call of this bird is most difficult to imitate correctly, and I have met with but one or two gunners who could successfully accomplish it. The attainment, however, appears to be of little service, as the bird is attracted quite as easily by the note of the Common Redshank, or the call of the Grey Plover.

Though the Spotted Redshank is occasionally met with in winter, I have never fallen in with the bird at that season.

The specimens in the case were shot on Breydon mudflats, one in August, 1871, and the other in August, 1872.

REDSHANK.—(Summer.)

Case 157.

This bird is common from Sussex to Caithness, being a permanent resident in the British Islands, though perhaps less numerous in the winter than at other seasons.

It breeds in marshes in various parts of the country, being particularly abundant, as might be expected, in the Broad districts of Norfolk. About August or September, when the young have gained sufficient strength, they leave their inland quarters and betake themselves to the sea-coast and salt-water mudbanks, joining occasionally in immense flocks.

On rough stormy nights, in the early part of the autumn, these birds are often attracted by the glare of the lights, and may be heard, in company with other Waders, flying over large towns, to the great astonishment of the inhabitants.

I have in two or three instances discovered eight eggs in the nest of the *C*ommon Redshank, it being in every case quite evident, from the difference in the colour of the eggs, that two birds had laid in the same nest. I have also seen seven and eight eggs in the nest of the Greenshank in Sutherland and Ross-shire; though in neither case could I distinguish the slightest difference in the eggs, the whole appearing to be one set ; this, however, was unlikely, as several other pairs of birds were observed at no great distance.

The specimens in the case were obtained in Glen-lyon, in Perthshire, in June, 1867.

THRUSH.

Case 158.

This well-known songster is widely dispersed over the British Islands,` the nature of the country, however, being less suited to its habits in the north; it is not so abundant in the wild districts of the Highlands.

Its neatly-built nest is placed in a variety of situations, at one time high in the branches of some lofty tree, at another among the ivy and other creeping plants on some overhanging bank, or even on the ground itself; furze-bushes, faggot-stacks, piles of old rubbish, ruined buildings, and indeed almost any site in which a nest could possibly be fixed, being occasionally resorted to.

By far the greater number of the Thrushes that make their appearance on the south coast during snow and frost, appear to feel the effects of the severe weather more readily than any other species, being frequently found moping, half-starved, under banks and hedges, while Blackbirds and Redwings are able to retain their usual condition.

I have on several occasions noticed Thrushes very abundant on the Bass Rock during autumn, being probably attracted to the spot by the number of snails among the old ruins.

Thrushes and Blackbirds, when feeding on the berries of the hawthorn in the winter, swallow them whole, and shortly afterwards cast up the stones.

The case is copied from a sketch taken in an old sawpit in the neighbourhood of Brighton.

The specimens were obtained at Portslade, in Sussex, in May, 1874.

MISSEL THRUSH.

Case 159.

Though not nearly so common as the preceding species, the Missel Thrush is well known from north to south.

The " Screech," as this bird is called by the natives in Sussex, is one of our earliest breeders, the young being frequently met with in the beginning of April.

The specimens in the case were obtained at Potter Heigham, in Norfolk, in April, 1871.

RING OUZEL.

Case 160.

The Ring Ouzel arrives in the spring, is widely distributed over the northern part of the country during the summer, and takes its departure early in the autumn. When living in the Highlands I have noticed these birds as being most destructive in the gardens, feeding greedily on currants, raspberries, and other fruits.

I once took a fine male in a trap set for a stoat, in the ruins of an old shealing in Perthshire. The bird must have had considerable difficulty in forcing its way up the narrow track that was left open, as we had removed several large stones, and then built up the trap and bait in the centre of the wall.

The specimens in the case were shot in Glenlyon, in Perthshire, in May, 1866.

REDWING.

Case 161.

The Redwing arrives in the north of Scotland early in the autumn, and gradually spreads itself over the country on the approach of cold weather, leaving our shores in the spring to rear its young in the north of Europe.

The specimens in the case were obtained in January, 1866, in the neighbourhood of Brighton.

REDSHANK.—(Autumn.)

Case 162.

The mature bird in autumn plumage, and the young of the year, are shown in the present case.

In September, 1873, I shot an immature bird of this species, that I mistook for a Spotted Redshank, while feeding on a mudbank in company with a large flock of other waders. On measuring its legs from the thigh downwards, it was exactly $1\frac{1}{8}$ inches longer than an old bird killed at the same discharge. I should certainly have preserved such a gigantic specimen, had it not been unfortunately plucked by mistake.

The mature bird was shot on Breydon mudflats, in the autumn of 1873, and the young on the shores of the Firth of Forth, in Haddingtonshire, in September, 1874.

SANDERLING.—(Autumn.)

Case 163.

The present case shows the immature Sanderling, on its first arrival in this country in the autumn.

The specimens were obtained on the Kentish coast, in September, 1869.

SANDERLING.—(Winter).

Case 164.

The Sanderlings in winter plumage were obtained on the south coast, between Rye harbour and Dungeness Point, in March, 1866.

SANDERLING.—(Summer.)

Case 165.

The Sanderling visits us in the spring, on the way to its distant breeding-grounds, returning again in the autumn, and occasionally remaining in considerable flocks throughout the winter.

The specimens in the case were shot on the shores of the Dornoch Firth, in June, 1869.

LARK.—(Summer.)

Case 166.

It may not be generally known that immense flocks of Larks arrive in this country during the autumn from the north of Europe.

I have fallen in with them daily about the middle of

October, when steaming in the North Sea many miles from land, flying direct for the Norfolk and Suffolk coast. They usually keep company in compact bodies of several hundreds, but now and then, a few, fatigued by the journey, would follow for a short distance, and then settle on board, where they would creep into the first quiet corner, and puffing themselves out like balls, would soon be at rest.

Larks are by far the most numerous of all the birds taken on board the light ships off the eastern coast, immense clouds being reported to have been often noticed hovering round the lamps during a drizzling rain. After these come the Starlings,* and then the Stormy Petrels. The latter, however, do not strike the lights, but settle on the vessels by day during protracted rough weather; they are then generally so thoroughly worn out, that if thrown up in the air they will immediately return on board. Gulls occasionally come in contact with the lamps, but it is by no means common for any species, except the Skua Gull, to be taken in that manner. I was informed by the mate of the *Newarp* that he had once found as many as three Skuas on deck during his watch, one of which, a large brown-coloured bird,† that he mistook in the dark for

---

* The first thousand wings that I received during the autumn of 1872 were made up as follows :—Larks, 520 ; Starlings, 348; Stormy Petrel, 45 ; Brown Linnet, 15 ; Greenfinch, 21 ; Brambling, 6 ; Fieldfare, 2 ; Forktailed Petrel, 1 ; Knot, 2 ; Blackbird, 20 ; Redwing, 13 ; Chaffinch, 15 ; Tree Sparrow, 3 ; Rook, 2 ; Snipe, 1 ; Kittiwake, 1. A few of the Warblers were taken the following spring, also one Swallow, but no Martins. I also received the wing of one Razor Bill.

† Probably the Great Skua.

a fowl, as it was lying disabled in a corner, inflicted a most severe bite on his hand. From all I could learn, the species usually taken were either immature Pomarine or Arctic.

Hawks and Owls are sometimes captured on board, but they are mostly observed before dawn, perched on some part of the rigging near the lamps.

On one occasion the glass of the lights of the *Newarp* was found broken, and a Duck of some kind, that the crew were unacquainted with, was discovered inside the lantern.

Grey Geese, Mallard, and various sorts of wild fowl used now and then to be obtained, but, from all I could hear, the numbers of these birds have diminished greatly of late years.

Snipe were stated to be among the commonest captures, but during the whole of the winter I only received a single specimen of this species. Woodcocks are supposed by the light-keepers to be particularly swift-flying birds, as they are generally picked up on deck much cut and injured by striking against the lamps or rigging. The real cause of such mishaps must, I should imagine, be attributed to the weighty condition of the bird at the time of the accident. I found that the Knot was well known as being of frequent occurrence. They were described as flying in large bodies, and numbers falling at once on deck. One of the men declared that the last flocks of these waders that he had observed, had suddenly shied when close to the lantern, and that but one or two had been taken.

It is stated that many years ago over one thousand birds were one morning collected on board the *Newarp*. Whether this was really the case or not I do not pretend to say,* but all my informants agreed in the fact that there was every year a great falling off in the number of birds so taken.

A fall of snow and a cold wind from the north-west is certain to bring enormous flights of Larks and other birds along the south coast; they all appear intent on making their way from east to west, occasionally passing in continued streams from daylight till dark. The Larks are usually the first birds to show; a slight covering of snow being sufficient to move them, while it requires a few days' continuance of severe weather before Fieldfares, Redwings, and Blackbirds appear in any numbers.

At such times the whole of the bird-catching fraternity of Brighton are engaged in the work of destruction. Should a strong cold wind from the north-west be blowing, the course of the birds is close to the ground, and thousands are captured in the nets. There is considerable competition for what are considered the best pitches, numbers of men leaving Brighton shortly after midnight, and depositing their packs on the ground they intend to occupy, to reserve the spot; they seek what shelter they can till daylight behind some bank or stack. From thirty to fifty dozen are commonly captured, and the takes not unfrequently reach as high as eighty dozen.

I have myself seen over 200 clap-nets at work on a

---

* I afterwards learned from an old man, who declared he was on board at the time, that the above statement was perfectly true. He also added that 600 of the birds, which were principally Larks, were put into one gigantic pie. I have a slight recollection of seeing an account of this fact in some paper, which published the history of the light-ships of the eastern coast.

favourable day, and as scores of drag-nets are out as soon as dusk sets in, some idea of the number of birds caught may be formed.

The whole of the specimens in the case were obtained in the Potter Heigham Marshes, in the east of Norfolk, in June, 1870 ; the Stoat being surprised while carrying off the young Lark.

GOLDEN ORIOLE.

Case 167.

The showy plumage of the Golden Oriole greatly interferes with its chance of a quiet life in the British Islands.

If not molested, it is most probable that these birds would soon become regular summer visitors to our shrubberies and gardens. I have myself seen the nest and eggs in Norfolk, and the fact has lately been recorded in the papers of a brood or two being reared in Kent.

The greater number, however, of those that make their appearance are speedily shot down ; and the reception that they meet with gives those that escape but small inducement to pay another visit to our shores.

The specimen in the case was shot between Shoreham and Lancing, in Sussex, in April, 1872.

FIELDFARE.

Case 168.

This is one of our most familiar winter visitors; the very name of Fieldfare seeming to be associated with snow and frost.

The specimens in the case were shot in the neighbourhood of Brighton, in January, 1866.

LARK.—(Immature.)

Case 169.

In the present case the young birds are shown in their nestling plumage.

The specimens were obtained between Shoreham and Lancing, in Sussex, in July, 1874.

RUFF.—(Summer.)

Case 170.

This singular bird is rapidly decreasing in number in the British Islands. Though several make their appearance every spring in the eastern counties, there are at the present time but two or three localities where they remain to rear their young; the swamps and marshes they formerly frequented being so reduced by the improved system of drainage, that few spots suitable to their habits are left.

On their first arrival, about the second week in April, the long feathers, which form the frill round the neck of the male, are but half grown. At this time they are seen (or rather used to be) in flocks of from ten or twenty to five or six times that number. On two or three occasions, in Norfolk, I have been able to crawl within a few yards of one of these large bodies, and have had a first-rate opportunity for observing their pugnacious habits. Their battles appear to be of but short duration. A couple of Ruffs square up to one another for a moment or two, and then separate to feed, or again go through the same performance with their nearest neighbour. Though they occasionally jump and strike after the manner of a game cock, I have never noticed any of the combatants to receive the slightest injury.

A large flock of Ruffs and Reeves is a most curious sight; the various plumages of the males, as they run rapidly here and there, giving a kind of piebald appearance to the whole assemblage.

The specimens in the case—male, female, and eggs —were obtained in the marshes in the neighbourhood of Potter Heigham, in Norfolk, in May, 1870.

PURPLE SANDPIPER.—(Summer.)
Case 171.

Though the nest of the Purple Sandpiper has never been discovered on our shores, the bird itself may be met with at all seasons. It is, perhaps, most abundant along the rocky coast between the Fern Islands and the Bass Rock, occasionally, however, making its appearance in considerable numbers all round the island, in every case where I have observed it confining itself to the sea-shore.

The specimens in the case were shot on the Carr Rocks, off the coast of Haddingtonshire, in June, 1865.

PURPLE SANDPIPER.—(Autumn.)

Case 172.

The specimens in the case were shot while resting on one of the breakwaters near the north pier in Yarmouth harbour, during the gale, in November, 1872.

RUFF.—(Autumn.)

Case 173.

The young in the immature plumage and old birds in their autumn or winter dress are still plentiful on the mudbanks on our eastern coast; the greater number, I expect, having arrived from foreign countries. I have once or twice met with the Ruff during winter, but I believe that their occurrence at that season is the exception and not the rule.

The specimens in the case were shot on **Breydon** mudflats, in August, 1871.

SHORE LARK.

Case 174.

The Shore Lark was formerly considered to be a rare bird in Great Britain, a few only having been met with during severe weather. At the present time, however, it is well known that there are several localities on the eastern coast where they may be always found from November to March. They are usually seen in small parties of from three or four to a dozen, or even a score, being easily approached, and appearing to be quite unsuspicious of danger.

The specimens in the case were obtained along the coast, between Blakeny and Salthouse, in Norfolk, in December, 1871.

BLACKBIRD.

Case 175.

The sketch from which the present case is copied was taken on the Bass Rock ; the nest having been placed in an old chimney, in the ruins of the Governor's house, which still stands among the fortifications. Seven or eight years ago a Blackbird's nest was built on the west side of the rock, within a few feet of the breeding-place of the Peregrine Falcon.

The specimens in the case were obtained in the neighbourhood of Brighton, in the spring of 1875.

STARLING.—(Maturd.)

Case 176.

Equally at home in town or country, the Starling is well known in every part of the British Islands.

In the flat districts of the eastern counties they collect during autumn in immense flocks, and early in the evening repair to the reed-beds, where they take up their quarters for the night. The extraordinary numbers that roost in such situations occasionally break down the reeds, and cause considerable loss to the owners of the beds.*

---

* The reed is extensively used for thatching and other purposes in the eastern counties.

The specimens in the case were shot at Offham Chalk Pit, near Lewes, in Sussex, in May, 1872.

WOOD LARK.

Case 177.

The Wood Lark shows itself occasionally in great numbers along the south coast when snow is on the ground. During the storm in January, 1866, I saw between six and seven dozen that had been captured by one bird-catcher alone, between Rottingdean and Newhaven.

They are to be found nesting in most counties in England, and occasionally (though I have never met with them myself) in the south of Scotland.

The specimens in the case were shot near Rottingdean, in Sussex, in January, 1866.

DOTTEREL.

Case 178.

The Dotterel appears unfortunately to visit our shores each year in rapidly decreasing numbers.

It arrives in April or the beginning of May, and after resting for a short time on the South Downs and other open spots, continues its journey to its breeding-grounds on the higher ranges of the north of England and the Highlands of Scotland ; again being seen occasionally in the south on its return in the autumn.

The specimens in the case were obtained on the hills to the north of the Lyon, in Perthshire, in June, 1866.

RING DOTTEREL.

Case 179.

The present species is common all round our coasts, being also occasionally found breeding on the banks of rivers and on sandy warrens in the interior of the country.

It is to be met with at all seasons, but perhaps most abundantly in the spring and autumn.

The specimens in the case were obtained in the neighbourhood of Rye, in Sussex, in the summer of 1862.

INTERMEDIATE RING DOTTEREL.

Case 180.

A diminutive form of the Ring Dotterel is here shown. It is easy to perceive that the birds are in every respect smaller than the specimens in the preceding case : the feathers on the back are of a darker shade, and the legs are also finer in form, and of a deeper shade of orange.

These birds usually make their appearance in flocks about the second week in May, at which time the larger variety are busy with their young broods.

I have never found these birds nesting in the British Islands, but mixed flocks of old and young are met with during the autumn.

The specimens in the case were shot between Shoreham and Worthing, on the Sussex coast; the mature birds being obtained in May, 1870, and the young in the following September.

KENTISH DOTTEREL.

Case 181.

The true home of this handsome little Plover in the British Islands is the flat line of coast that lies between Rye Harbour and Dungeness Point; here it is found in numbers during the spring and summer months, usually departing about the beginning or middle of September. It may be met with at a few other spots along the south coast, and also occasionally in Norfolk and Suffolk, but its appearance in those localities is very uncertain.

The specimens in the case were shot on the sands between Rye and Lydd, on the shores of the English Channel, early in May, 1866.

COMMON BUNTING.

Case 182.

The Common Bunting is of frequent occurrence in the British Islands, being found, perhaps, more abundant in Sussex than any other county.

The specimens in the case were shot in the neighbourhood of Brighton, in May, 1872.

STARLING.—(Immature.)

Case 183.

The young are here shown in their nestling plumage, one or two of the specimens exhibiting a few feathers of their first moult.

They were obtained at Portslade, near Brighton, in the summer of 1872.

SPOTTED FLYCATCHER.

Case 184.

This is one of the latest of our summer visitors, seldom appearing till well on in May, then, rapidly dispersing over the country, it rears its young, and takes its departure early in the autumn.

The specimens in the case were obtained at Portslade, near Brighton, in the summer of 1874.

SNOW BUNTING.

Case 185.

Though usually supposed to leave our shores on the approach of summer, a few pairs of Snow Buntings remain, and rear their young in some of the wildest districts of the Highlands of Scotland.

Early in the autumn immense flocks make their appearance in the north, and gradually work round the coast, till hundreds are often observed in severe weather in the immediate vicinity of the English Channel.

Their numbers, in Sussex, vary with the seasons, but few being noticed in mild winters.

The specimens in the case were shot on the South Denes, near Yarmouth, in Norfolk, in November, 1872.

SNIPE.—(SUMMER.)

Case 186.

The Snipe is a permanent resident in Great Britain, breeding wherever suitable localities are met with from north to south. Its numbers are also considerably increased in the autumn by arrivals from the Continent. It is probable that two or even three broods are occasionally reared during the season.

I have seen young birds as early as the beginning of April, when snow was on the ground, and the nestlings in the case were taken as late as the 9th of July, having been only hatched on the previous day.

The male and female were shot on the Fendom, near Tain, in Ross-shire, in the spring of 1869 ; the young, as stated above, being procured on the same ground in July.

WOODCOCK.—(Summer.)

Case 187.

The Woodcock breeds abundantly in the south of England ; and there are, indeed, but few counties in which the bird is not occasionally seen during the summer months, though the nest itself may escape observation.

Its curious habit of carrying its young has given rise to innumerable discussions in the natural history columns of the sporting papers, each writer asserting that the operation was performed in a different manner. There is, however, but little doubt that the young bird is firmly pressed between the thighs of the parent, and so transported from one spot to another.

In addition to those bred in this country, large flights arrive from the north of Europe during the autumn, the birds beings occasionally found in a very exhausted condition.

The case is copied from a sketch taken in the Tarlogie Woods, near Tain, in Ross-shire, at which place the female and eggs were obtained in June, 1869.

WOODCOCK.—(Winter.)

Case 188.

As will be readily seen by the specimens in the present case, the plumage of the Woodcock in winter is considerably darker than in summer.

· The birds were shot in Glenlyon, in Perthshire, in November, 1875.

SNIPE.—(Winter.)

Case 189.

While punt-gunning one winter on a river in the north of Scotland during severe frost, I noticed that Snipes were collected in numbers along the banks where the mud was kept soft by the action of the tide. · As a novel proceeding I tried one shot at them with the big gun, but the poor birds were so tame that it could hardly be considered sport, and fowl being plentiful on the water at the time, I left them alone, in hopes of renewing · the acquaintance on some future day. I, however, discovered when the weather changed that I had lost my chance, as after the breaking up of the frost not a snipe could be found within a mile of the spot.

I shall not readily forget a rather amusing incident that occurred that evening.

I was stopping at a first-rate hotel, which, as is

commonly the case in the Highlands, was situated in a remarkably wild and, during winter, deserted region.

Having finished a capital dinner, and being tired of my own company, I strolled into the kitchen to see what was going on. Here I found the punt-gun propped up on a couple of chairs in front of a roaring peat fire, with two or three keepers sitting smoking beside it. As I noticed that several sparks from a lump of peat with which one of the men was lighting his pipe fell over the lock, I inquired if the charge had yet been drawn. " Yes," remarked " John," the punt-man, with the air of one who thoroughly understands his business and has properly performed it; " she's washed out, loaded, and carefully primed ready for the morning." On inspecting the lock I discovered that the covering to the nipple was simply a piece of brown paper such as Highlanders use for tinder, and consequently extremely liable to ignite from a spark ; so I suggested that the gun should be placed in the far corner of the room, where it would still be protected from the frost.*

Nothing, however, would satisfy the landlady, who came in at this point, but the immediate removal of the dangerous weapon. I accordingly ordered the men to take the gun out, and draw the charge.

In less than two minutes there was a deafening explosion, followed by a fearful crash : the glass was blown in, the lights blown out, the landlady fainted, the lassies screamed, and the dogs barked. On rushing out to see what had happened, I learned that after

---

* Some gunners have an idea that it is dangerous to allow the frost to get into the barrel of a punt-gun.

cleaning out and loading the gun, the men had carried the rods down to the punts when they had gone to see that all was snug for the night, and consequently had no means at hand for drawing the charge; so, placing a cap on the nipple, they had steadied the butt on the bricks of the yard, and holding the muzzle in the air, had fired the charge, with the result described. On subsequent inquiry I found out that " John," who had rather a spite against the landlady (as that stern matron had reprimanded him for some unbecoming levity she had detected between him and one of the damsels of her establishment) had done it in hopes of giving her a fright, without having bestowed a thought on the panes of glass that would be blown in by the concussion, and—naturally fell to my share to pay for.

This is one of the few birds that I have seen in the act of striking the wires of the telegraph. While the train was entering the station at Tain, in Ross-shire, one evening in March, 1869, I noticed a snipe spring from the side of the line, and, rising straight in the air, come in contact with the wire, and immediately fall disabled to the ground. As I was returning from shooting in the neighbourhood I had my gun with me, and on proceeding to the spot I bagged, in the few minutes of daylight that were left, four and a half couple of Snipe and an immature Golden Eye. The greater part of the birds rose from a small brick drain of water that ran from the station, two or three escaping through flying off in line with the telegraph wires, as I was afraid to fire lest some damage (the station-master being present) might be laid to my charge.

The specimens in the case were obtained in Pevensey

Levels, in the winter of 1866; here, formerly, a good bag of long bills might usually be made during autumn, winter, and early spring. For the last ten or twelve years, however, their numbers have been gradually falling off, till, at the present time, I am afraid the ground is almost useless as a Snipe beat.

YELLOW BUNTING.

Case 190.

The Yellowhammer is one of our commonest birds, being plentiful at all times and seasons.

I once observed a large number of these birds, together with Chaffinches, feeding on the flesh of a dead horse which was hanging against a dog kennel in the Highlands during a heavy snowstorm. On the joints being placed on the ground and the snow swept from them, the poor birds came down in hundreds and settled on the meat.

They appear to be more hardy than our northern visitors, the Bramblings, as during severe weather the large flocks of that bird disappear almost entirely from the Highlands, and are found scattered over the southern counties of England, while the Yellowhammers contrive to weather the storm in their native glens, obtaining what food and shelter they can round the farm-buildings and houses.

The specimens in the case were shot in the neighbourhood of Brighton, in the summer of 1872.

PIED FLYCATCHER.
Case 191.

Though occasionally seen while on its passage in spring and autumn, I have never met with an instance of this species remaining to breed in the south of England. It, however, appears to be far from uncommon during the summer months in some of the more wooded parts of Cumberland, and probably spreads into a few of the adjoining counties.

In the spring of 1867 I shot a female on the Bass Rock.

The specimens in the case were obtained among the fine old timber in the park at Edenhall, in Cumberland, in June, 1876.

C R E E P E R.
Case 192.

This active little bird, which is apparently always in motion, running up and round the stems of the trees, is found abundantly from north to south.

Its nest is placed in holes in trees, buildings, and old walls.

The present case is copied from a sketch taken near Aberfeldy, in Perthshire, at which place the nest and eggs were also obtained.

The birds themselves were shot near Plumpton, in Sussex, in March, 1866.

CIRL BUNTING.
Case 193.

Though not uncommon in several of the midland counties of England, I have only met with the present species in Sussex, where it is by no means scarce in that portion of the county that lies within about ten or twelve miles of the Channel.

During winter they join in flocks, at times associating with other small birds, such as Yellowhammers and Chaffinches.

The specimens in the case were obtained in the neighbourhood of Brighton, in November, 1873.

REED BUNTING.

Case 194.

The Reed Sparrow, as this bird is more commonly called, is most abundant in marshy districts, being especially numerous in the locality of the Broads of Norfolk and Suffolk. It is, however, to be met with as a resident in almost every county in England and Scotland.

The specimens in the case were obtained in the east of Norfolk, in the summer of 1871.

JACK SNIPE.

Case 195.

The Jack Snipe is only a winter visitor, arriving in September and taking its departure in April.

The erratic flight of this curious little bird is often a puzzle to nervous shooters; many a charge of shot being wasted before it is discovered that by waiting a few moments an easy chance is obtained.

The specimens in the case were shot in Pevensey Marsh, in March, 1866.

TUFTED DUCK.

Case 196.

On the lochs in the Highlands, and on the lakes and broads in England, the Tufted Duck may be found in smaller or larger flocks during the winter months.

It is not till the latter end of the season that any except the oldest drakes assume their handsome plumage. On their first arrival in the autumn, the whole of the birds composing the flocks present the appearance of females or young.

The males with the long tuft are supposed by the gunners, in the east of Norfolk, to be in no way related to the rest of their families, and, together with the drake Golden Eyes, are termed " Old Hardweathers." I happened one night while flight shooting to kill a perfect specimen of both species with the same barrel, and nothing would convince my punt-man that they were not male and female of one and the same kind.

Like the Scaup and Pochard, with which they occasionally associate, the Tufted Duck is an excellent diver. A successful shot with the punt gun at a large flock is certain to produce a number of cripples ; these it is almost useless to pursue if there is the slightest ripple on the water, as each makes off in a separate direction, only occasionally showing the point of its bill for air.

One of the drakes in the case, while attempting to escape by diving, was caught round the neck by a weed and drowned. The water being at the time of the occurrence as clear as glass, I noticed some bubbles coming to the surface, and could plainly see him at a depth of three or four feet struggling with his head held down as if in a noose. In a few moments he was quiet, hanging like a criminal (only in the inverse manner), suspended by the neck.

The specimens in the case were obtained on the broads in the east of Norfolk, early in 1873.

COMMON TERN.—(Mature.)
Case 197.

During fine still weather, early in May, the first arrivals of these birds may be looked for. Their breeding-stations, which are still (though rapidly decreasing) numerous in many parts of Great Britain, present a most animated appearance by the beginning of July. Young birds of every age and stage may then be seen, together with the old ones, busily attending to their wants; the whole group affording a sight both interesting and amusing.

Though occasional stragglers may be met with as late as November, by far the greater number of these Terns have taken their departure for a warmer climate by the middle of October.

The specimens in the case were obtained on the shores of the Dornoch Firth, in Ross-shire, in June and July, 1869.

ROCK DOVE.
Case 198.

All round the coast of Scotland and its adjacent islands, wherever rocky caves are met with, the present species is almost certain to be found as a tenant. They are occasionally stated to have been seen frequenting the chalk pits and other similar situations in the south of England, but in every case I have discovered the bird to be the Stock Dove. In some parts of the north I have seen Wild Pigeons, of sandy and other light colours, living among the rocks as wild and untamed as the present species.

The young were taken from a cave on Longa Island, off the west coast of Ross-shire, in May, 1868, the old birds being killed at the Cromarty Rocks, in June, 1869.

The case is copied from a sketch taken of the spot where the young were procured.

STOCK DOVE.

Case 199.

The Stock Dove is common in the south of England, and I have repeatedly observed them in large flocks in the east of Norfolk, feeding on the pea-fields, during the summer months.

They appear to breed in a variety of situations, holes in old timber being, as most authors state, their favourite nesting-place; they, however, occasionally rear their young on the branches of a tree, like the Ring Dove, at times in a rabbit burrow, and also in the face of a cliff.

The young birds (or squabs, as I believe juveniles of the Pigeon tribe ought properly to be termed) were taken from a nest in every respect resembling that of a Wood Pigeon, near the top of a small spruce fir tree of about thirty feet in height, in the neighbourhood of Brighton, in June, 1874; the male and female being shot in the Potter Heigham marshes, in the east of Norfolk, a year later.

The case is copied from a Stock Dove's nest in an elm-tree at Falmer, near Brighton.

COMMON TERN.—(IMMATURE.)
Case 200.

The Common Tern is here shown in the plumage of the first autumn.

The specimens were shot at sea, off the Suffolk coast, in August, 1873.

GOLDEN EYE.—(MATURE.)
Case 201.

This handsome Duck is common along most parts of the Scotch coast, frequenting the firths and also the inland lochs. In England, however, except in severe winters, it is by no means so abundant.

Young birds and females may be met with at all times from the beginning of October till April, but it is seldom that the mature Drake is observed till January or February.

This species is generally found in pairs, or small parties of from half a dozen to double, or perhaps three times that number. The females and young are quite unsuspicious of danger, and when feeding can usually be approached with but little manœuvering. The old drake, however, is one of the most wary of wildfowl, taking wing on the slightest signs of danger.

The specimens in the case were shot on Loch Slyn, in the east of Ross-shire, in March, 1869.

GOLDEN EYE.—(Immature.)

Case 202.

The immature males of this species are observed, during their first winter, in the plumage of the specimens in the case.

It is not often that the Golden Eye is seen on land · the two birds in the case, however, were shot as they rose from the bank of one of the islands on Hickling Broad, in January, 1873.

WHITE WING TERN.

Case 203.

The present species is but an accidental visitor to our shores, those previously obtained having been for the most part met with during spring and summer, in the eastern counties.

No recorded instance of the immature bird occurring in this country has come under my observation, though I happened to see a fine specimen which was shot on Horsey Mere, in a private collection in that neighbourhood.

At the time the birds in the case were killed, I had but little opportunity for observing their habits. I first caught sight of them while fishing in the channels on Breydon flats, before it was fairly light, during a heavy storm. After following them some time they all settled on a mudbank, appearing to be attracted by a pair or two of fine old Grey Plovers which were resting there, and, happening to alight at the moment I came within gunshot, the whole were obtained at one discharge.

In the grey of the morning I at first mistook them for Black Terns, and was surprised to notice them pitching in the water for food like the *Common* or Arctic Tern; the Black usually feeding on insects, which it captures over the water, in the same manner as the Sand Martin.

The birds were shot after a most tempestuous night, early in the morning of the 26th of May, 1871, on Breydon mudflats.

TURTLE DOVE.

Case 204.

The Turtle Dove is only a summer visitor to our shores, arriving in May and departing early in the autumn.

Though this bird is occasionally observed in the northern parts of the island, it is most numerous in the southern counties, being particularly abundant in Sussex.

The specimens in the case were shot at Portslade, near Brighton, during the summer of 1872.

RING DOVE.

Case 205.

Though the present species is a resident in all parts of the British Isles, the immense flocks that are at times met with would lead to the belief that their numbers occasionally receive additions from the *Continent*.

The farmers in various parts of the country complain of the depredations committed by these birds, and frequently organise societies for their extermination; but whether their efforts will ever be attended with success appears to be extremely doubtful, one fact at present alone being certain, viz., that their attempts to thin them down have hitherto been perfectly unavailing.

As to hazard a guess at anything approaching the numbers that occasionally congregate together would appear incredible to those that have never had an opportunity of observing them, it will be sufficient to state, that in favourable localities they often join in flocks of several thousands.

The specimens in the case were obtained in the neighbourhood of Brighton in the spring of 1870.

WHITE WING TERN.
Case 206.

The five Terns in the present case formed part of a flock of seven that I met with on Hickling Broad.

They were first observed hawking for flies over the water in company with the Sand Martins, and, having previously obtained as many as I required as specimens, I had a good opportunity of watching their habits.

They appeared quite fearless, occasionally approach. ing and hovering within a few feet of the boat; though the water was shallow, and small fry abundant, they never attempted to capture a single fish, frequenting only those parts of the Broad where the Sand Martins were seen, both species taking their food in precisely the same manner.

Towards dusk, as I found the Martins were leaving, I rapidly procured the Terns, the sixth falling dead to a long shot in the middle of a reed-bed where it

M 2

was useless to search; and the seventh, after having for some minutes complacently watched the slaughter of his companions, took his departure without offering a chance.

On the following days there were again small flocks on the Broads, those seen latest appearing to be younger birds, their breasts being strongly marked with white.* While the weather was cold and windy they pursued their course straight away to the north-east; those, however, that were seen, when the sun had brought out the insects and Sand Martins, remained for some hours hawking over the water.

I learned from the keeper that on three or four consecutive days before I fell in with the first flight he had observed some small parties of "Daws"† that he never remembered to have seen before. These were probably birds of the same species.

Three were also noticed on Breydon mud-flats, about the same time, by one of the gunners who had seen the birds I had obtainedthere a couple of years previously.

It would appear that several parties of these Terns must have continued passing across the east of Norfolk for about a week during the latter end of the month.

The specimens in the case were shot on Hickling Broad on the 28th of May, 1873.

---

* In the immature bird the breast is white.

† The natives of this locality style the whole of the Tern family "Daws;" the common or Arctic being the "White," and the Black the "Blue Daw."

SCOTER.

Case 207.

The Scoter is common at most seasons in flocks off the greater portion of our coasts.

A few remain to breed in the northern counties of Scotland, though by far the larger number of our visitors are reared in the far north.

The specimens in the case were obtained in Strathmore, in Caithness, in June, 1869.

VELVET SCOTER.

Case 208.

A few of these fine Ducks are occasionally seen in the English Channel during winter; the numbers, however, that are met with in the south have fallen off greatly of late years. From October till April they may be still observed in large flocks in the firths of Scotland and the northern islands.

The specimens in the case were shot at sea off Hastings, in January, 1860.

BLACK TERN.—(Summer.)

Case 209.

Though formerly nesting in the eastern counties the Black Tern is at the present time only a visitor to our shores in spring and autumn. This species is usually the first of the Terns to appear in the spring, the earliest arrivals being frequently noticed in the beginning of April; I have, however, observed them passing all through May, and occasionally as late as the second week in June.

The specimens in the case were obtained on Heigham Sounds, in the east of Norfolk, in May, 1871.

Case 210.

Case 211.

BLACK TERN.—(Autumn.)

Case 212.

The old birds showing the change into winter plumage, and the young in their first feathers, are occasionally observed in this country as early as July; but it is not till August and September that we are visited by the main body while on their way to their winter quarters.

The specimens in the case were shot on Hickling Broad in the beginning of August, 1873.

LONG-TAILED DUCK.

Case 213.

This Duck is only a winter visitor to the British islands, arriving in September, and departing in March or April; immature or backward birds being, however, occasionally observed in May and June.

Though a few straggling parties at times find their way as far south as the English Channel, this bird is seldom met with in any numbers except off the coast of the north of Scotland.

Its note is most peculiar, one of its local names, viz., "Coal and candle light," being derived from a resemblance it is supposed to have to those words, which the bird pronounces in a sing-song manner.

The specimens in the case were shot at the mouth of the Little Ferry, near Golspie, in Sutherland, in March, 1869.

RAZOR BILL.—(Summer.)

Case 214.

The birds are here represented at the foot of the Bass Rock.

Numbers of Guillemots, Puffins, and Razor Bills, may be observed about daybreak resting on the lower ledges, but the approach of a boat always drives them into the water.

The ordinary visitor to the rock would never imagine the animated appearance of the spot if viewed shortly after sunrise.

Both at the Bass Rock and the Fern Islands these birds, from some unknown cause, are rapidly becoming scarcer year by year. Indeed, I expect that they have now entirely ceased to breed on the Ferns. At the time of my last visit, in 1867, there was but a single pair frequenting the Islands.

At the more northern stations their numbers show no signs of diminution.

The specimens in the case were obtained in the Firth of Forth, in the vicinity of the Bass Rock, in June, 1865.

GLAUCUS GULL.—(Immature.)

Case 215.

Immature birds of this species are common along the north-east coast at most seasons of the year.

The mature Glaucus is, however, rarely met with except in severe weather, and it is even then seldom that they are obtained in the south. Though I have observed a few on different parts of the coast, I have never had a chance to procure a single specimen.

The bird in the case was shot while flying over the Hickling Marshes in the east of Norfolk, in December, 1873.

RAZOR BILL.—(Winter.)

Case 216.

The Razor Bill in winter is met with in numbers all round our coasts, a few miles off the land, following the shoals of sprats and other fish on which they feed. During protracted rough weather they occasionally suffer severely from hunger, those obtained after a storm being usually in poor condition.

The fishermen on some parts of the south coast appear to have a fancy for these birds. I have often been asked for any I did not want, the men declaring they made a capital dish when stuffed with onions.

The specimens in the case were shot in the English Channel during the winter of 1870.

LESSER BLACK-BACK GULL.—(Summer.)

Case 217.

The present species is perhaps the most numerous of our British Sea Gulls, breeding on various rocky parts of the coast, and in colonies of smaller or larger size on the inland lochs of Scotland.

The islands on Loch Maree, in the west of Ross-shire, are resorted to every season by thousands of pairs of these birds. They are here permitted to rear their young in comparative peace, as boats are (or rather were, for I have not visited the spot for some years) scarce on the loch, and it is but seldom that the country people are able to reach the islands to obtain their eggs.

Those who have only viewed this beautiful loch under the influence of a bright sky and a gentle breeze, would never credit the fury of the squalls that at times gather among the surrounding hills and burst with but scanty warning over its surface. On one occasion, when I had sent on the previous day to the keeper to ask for the use of the boat, I found on arriving at the spot that three girls had come down from the hills in hopes of getting out to procure a few basketfuls of eggs. After landing them on the Islands where the Gulls were most plentiful, we proceeded to search for the nests of Geese and Divers, or other rarities that might fall in our way. Though the early morning had been fine and still, the day by noon had clouded over and rain and wind set in. For some hours we delayed our return-voyage in expectation that the weather would moderate, but the longer we waited the worse grew the storm. At last, while attempting to reach the shore with a large cargo of eggs, we were struck by a squall, which came roaring across the loch, with a blinding cloud of spray, and driven back on one of the islands, the breaking of an oar sending two of the crew to the bottom of the boat, where they rolled about with the eggs, which were now being dashed from side to side. On working our craft into a sheltered bay, and landing our terror-stricken passengers, we were forced to work hard to repair the damages we had received, and by the time our defects were made good, the storm had abated sufficiently for us to make a second attempt.

The girls, who had crouched at the bottom of the boat, presented a most ludicrous appearance, being drenched to the skin with a mixture that resembled egg-flip, the whole of their spoil, consisting of several hundred gulls' eggs, having been smashed and beaten up into a kind of custard with the water that had broken on board. Some meat and drink, and the attentions of two or three sturdy keepers, eventually put fresh life into the disconsolate maidens; but when they took their leave in the gloaming, it was hard to recognise in the three bedraggled tramps the bright-looking lassies that had met us in the morning.

There is no doubt but that this species is very destructive to game and their eggs. I have during spring often trapped them on the moors in Perthshire, using as a bait either eggs or flesh. In Ross-shire, Sutherland, and Caithness, where their numbers are far greater, the loss that they cause to the game preservers must be very considerable.

At the time of the bringing out of the Sea Bird Act it was stated that gulls were of great assistance to the fishermen, by showing them the position of the fish, and so guiding them to the best spots for shooting their nets. This all looks very pretty and interesting in print, but I have yet to learn that the Sea-gull is a favourite with the fishermen.

When the shoals of mackerel arrive off the south coast in the spring of the year, scores of boats are engaged in watching for the fish to come to the surface; they then row rapidly to the spot, and shooting a net round them, frequently enclose large numbers. Should

any Gulls, however, be near at hand, their sharp eyes
are sure to detect the first ripple on the water, and
dashing down into the middle of the shoal, they drive
the fish to the bottom, and the men who may have
rowed hard for half a mile or more, and possibly paid
out a portion of their net, find their time and labour
thrown away, while the mischievous bird, with a de-
risive scream, sails off to repeat the performance at
the earliest opportunity.

While watching the proceedings I have often been
requested to kill the Gulls, the men declaring that,
what with the Bird Act and the gun license, they were
unable to help themselves, being forced to stand
quietly by while the birds snatched the bread from their
mouths.

The number of these Gulls that congregate in the
North Sea during the herring season in the autumn
is something enormous; here, again, they cause great
loss to the fishermen. I have been assured by the
masters of some of the luggers that they have
frequently been deprived of a last of herrings, and
occasionally up to even four or five times that quantity,
by their depredations.

As a last is over ten thousand fish, the number
might seem incredible to those who have never had
an opportunity of watching a large flock of these birds
gathered round a boat that is making a good haul.

The number that they swallow is small compared
with those they bite and shake from the nets. I have
myself repeatedly observed as many as a thousand or
two of the larger species of Gulls attacking the nets of
a single boat; at times, taking hold of the lines in their
beaks, they rise in the air, and attempt to shake out the

fish. The small boat is occasionally sent to drive off the birds, but if disturbed from one part of the nets, they rapidly commence operations on another.

After the boats have finished hauling, the birds are usually satisfied, and rest in large bodies upon the water for the remainder of the day. It is by no means uncommon to meet with a flock that extends a mile or two in length.

I have noticed that when any disease has broken out, and destroyed the fish on the fresh-water broads in the eastern counties, that the gulls rapidly become aware of the fact, and resort daily in thousands to the spot, feeding greedily on the decomposing remains that are floating round the banks.

The specimens in the case were obtained on the Bass Rock in June, 1867.

LESSER BLACK BACK GULL.—(IMMATURE.)
Case 218.

The specimens in the present case represent the bird in its various immature stages.

They were shot at sea, off the Norfolk coast, in the autumn of 1872.

HOOPER, OR WILD SWAN.
Case 219.

It is only in the wilder and more remote parts of our Islands that these birds can now be regularly looked for. Improvements in drainage and the increase of gunners are gradually lessening the numbers of our visitors, still, when suitable weather occurs, a few are sure to be seen in the neighbourhood of their former haunts.

GREAT NORTHERN DIVER.—(IMMATURE.)
Case 220.

During autumn and winter the immature birds of this species are commonly met with in various parts of the British Islands, frequenting both fresh and salt water.

The specimens in the case were shot on Hickling Broad in January, 1872.

ARCTIC SKUA.—(MATURE.)
Case 221.

This bird may still be found breeding in many parts of the north of Scotland and the adjacent islands, the nest being placed on the open moor.

On land, as at sea, the Arctic Skua for the most part procures its food by robbery; those that I have seen in Strathmore usually persecuting the unfortunate common Gulls that have the misfortune to nest in the same locality.

The male and female in the case were discovered, on being opened, each to contain four Smolts, or young Salmon. Fish, as a rule, is their diet, but this they occasionally vary with eggs, swallowing, I believe, the whole or the greater part of the shell, as I have often noticed castings composed entirely of egg-shells on the mounds where these birds are in the habit of resting

The specimens were obtained in Strathmore, in Caithness, in June, 1869.

KITTIWAKE.—(Summer.)

Case 222.

The Kittiwake is common all round the British Islands, breeding in the rocks that overhang the sea on numerous parts of our coasts.

We have few Gulls so thoroughly marine in their habits, the present species being seldom, if ever, met with inland.

Though but a comparatively small bird, the quantity of fish it is able to consume is perfectly astonishing. One of the specimens in the case, on being lifted into the boat, disgorged three large herrings; these could only have been swallowed a few minutes previously, being all as bright as silver.

To this poor, persecuted wretch, the " Sea-Bird Preservation Act " has certainly been a blessing, the senseless slaughter that took place round their breeding-stations every summer having been allowed to continue too long without interference.

The specimens in the case were obtained at the Bass Rock, in June, 1867.

KITTIWAKE.—(Winter.)

Case 223.

The mature bird in winter plumage, and the immature in their first feathers, are here shown.

During autumn and winter they are occasionally seen in immense flocks, following the shoals of sprats and other fish in the English Channel.

The specimens in the case were shot a few miles off Brighton, in the winter of 1870.

ARCTIC SKUA.—(Mature, Autumn.)

Case 224.

The present case shows a variety of shades and colours in the plumage of this singular species.

The perfectly black specimen on the left, together with the one immediately beyond it, and the sitting bird on the right, are females, the remaining four being males. It will be thus seen that there is no rule for the colouring of either sex.

In the autumn these birds are very numerous off the northern coast wherever Kittiwakes are plentiful. When the boats are hauling their long lines for haddies and whiting, hundreds of Gulls are attracted to the spot for the fish that fall from the hook while being lifted on board; these they snatch up within a foot or two of the boat, but are frequently forced to disgorge, should a Skua be near at hand. The robber appears to take no notice of the Gull, if sitting on the water, beyond watching it intently; but the moment it rises on wing he attacks it.

Three of the specimens in the case were killed by a double shot while swimming close to an unfortunate Kittiwake, which, having made a good meal, was either too full or too frightened to fly.

The birds were obtained in the Firth of Forth, in August, 1874.

BLACK-THROATED DIVER.—(Summer.)

Case 225.

Though but rarely seen in the south, this handsome bird is still abundant in several parts of the Highlands;

it appears to be more partial to the lochs that lie in wild, hilly districts, preferring those with rocky islands, on which it rears its young.

I have occasionally seen as many as from fifteen to twenty fine old birds at one time on a single loch during the summer months, diving, splashing, and screaming, and now and then pursuing one another, both above and below the surface. Whether these were birds that had been robbed of their eggs or young, or had only gathered together for the sake of company, I can form no idea, as possibly on my next visit to the same loch, but a single pair would be visible.

At times, when crossing the hills, I have noticed as many as eight or ten flying together; on such occasions they keep in a straight line, at regular intervals, one behind the other.

The newly-hatched young are covered with black down; like all waterfowl, they take to their natural element as soon as they leave the shell.

The specimens in the case were obtained in the west of Ross-shire, in May, 1868.

BLACK-THROATED DIVER.—(Immature.)

Case 226.

Immature birds of this species are not unfrequently met with during the winter in the south of England, at times being found on both fresh and salt water.

The specimen in the case was shot on Heigham Sounds, in Norfolk, in February, 1871.

ARCTIC SKUA.—(Immature.)

Case 227.

The barred specimen in the present case is probably in its second year; the two dark birds are, without doubt, but three or four months' old.

They were shot at sea, off the Bass Rock, in September, 1874.

HERRING GULL.—(Summer.)

Case 228.

From my own observations, I should be of opinion that the farmer rather than the game-preserver would have a right to complain of the damage caused by the present species.

I have never seen a single specimen captured in a vermin-trap set for the destruction of the other Gulls, nor have I ever observed them preying on either young game or eggs.

In the north they appear to prefer the cultivated tracts of land in the neighbourhood of the coast, where, after feeding in large flocks on the fields, they retire to the rocks to rest.

Gulls, when alarmed (as the Skua is well aware) usually vomit the contents of their stomachs. By firing a shot amongst a flock while sitting on the shore after feeding, and causing them suddenly to take flight, I have repeatedly found that some cast up a quantity of grain, and others large lumps of mussel-shells, which they uppeared to have swallowed whole.

That they can, however, make a meal of young birds

I have good proof, as some that I keep in confinement devoured a couple of young blackbirds that escaped from their cage, and also managed to bolt a Water Rail, to say nothing of numerous sparrows that they contrive to capture while feeding on their corn.

Some years ago these birds were plentiful on the Bass, where they nested every season.

The person who hired the rock, finding that the Jackdaws which had recently taken up their quarters in the rabbit-burrows near the summit were very destructive to the eggs of the sea-fowl, endeavoured to destroy them by laying down poisoned bread and butter; this, however, was greedily devoured by the larger species of Gulls, who suffered in consequence, and since that time there have been but two or three pairs of either Herring Gulls or Lesser Black Backs about the rock.

The case is copied from a drawing made on the north side of the Bass; the specimens, together with their nest and eggs, being obtained on the rock in June, 1867.

HERRING GULL.—(Winter.)
Case 229.

This case shows the mature bird in winter plumage, and the young in the second or third year.

The specimens were shot in Yarmouth Roads, in November, 1872.

BUFFONS' SKUA.
Case 230.

I have had but little opportunity for observing the habits of the present species.

A pair in immature plumage, which I noticed several times one day flying together in the early part of May, in the Channel, and the specimen in the case being all that I have met with.

The bird was shot close to the West Pier. at Brighton, where it had remained sitting on the water for several hours, being probably worn out by a continuance of stormy weather. It was obtained in November, 1870.

RED-THROATED DIVER.—(SUMMER.)

Case 231.

The Red-throated Diver is common during the nesting season in the north of Scotland, and in many of the adjacent islands. This species appears to be more plentiful where the country is flat, with small marshy pieces of water, than in the hill lochs, which are the true home of its relative the Black Throat. It is consequently most numerous among the "floes" * which abound in the central part of Caithness.

To state that the note of this bird is pleasing to the ear would scarcely be correct. I have seldom heard anything more melancholy than its dismal cries, which are frequently repeated both before and during the continuance of rough and stormy weather.

The specimens in the case were obtained in Strathmore, in Caithness, in June, 1868.

---

* Flat tracts of moor, with still, deep, black pools of water.

RED-THROATED DIVER.—(WINTER.)

Case 232.

During the winter months these birds are common all round our coasts, great numbers, in addition to our regular residents, arriving from the north of Europe, and leaving again on the approach of spring.

The specimens in the case were shot a few miles off Brighton, in December, 1870.

POMERINE SKUA.

Case 233.

Though occasionally compelling the Kittiwake to provide it with food, this Skua more commonly attacks the larger species of Gulls.

Immature specimens may be observed in numbers in the North Sea during autumn, following the large flocks of Lesser Black Backs always in attendance on the herring fleet. As winter advances they work their way south, and are occasionally met with in the English Channel.

I have at times seen a few fine old birds in the early part of the autumn off the Scotch coast, but the majority are without the long feathers in the tail.

On one occasion when at Yarmouth during the herring season I was told by the master of one of the fishing luggers that he had shot a bird or two while on the North Sea that he wished to show me. As I was aware he had been afloat for at least a fortnight I thought they might possibly be getting unpleasant, but being assured they were as fresh as when they were killed, I told him to bring them down for me to look at in the evening. The specimens, which proved to be an immature Gannet, a Pomerine Skua, and a bird I could not determine, stunk in such a horrible and fearful manner, that I requested him to take them away and let me see them by daylight, when I hoped to be able to examine the stranger more closely out of doors.*

On arriving at his house on the following morning, I found that they had been sold by his wife for sixpence to a musician, as an old and two young Hansers,† and were at that moment being cooked for the Sunday's blow-out, much to the disgust of his neighbours in the row, who were almost poisoned by the stink.

The specimen in the case was shot in the Firth of Forth, in August, 1874.

GREAT BLACK-BACKED GULL.—(Summer.)

Case 234.

This fine bird is to be met with all round the British islands, breeding in the rocks that overhang the sea, and also on the islands in several of the Scotch lochs.

During the summer months these Gulls are most destructive to the young of any game or waterfowl that fall in their way. I have more than once seen them rise from the carcase of a dead sheep, and have been assured by shepherds that they not unfrequently kill the sickly lambs, and at times even the ewes, if found in a weakly condition.

---

* The bird was about the same size and make, though slightly smaller, than the Pomerine, of a pale chocolate, with round white spots as large as a threepenny-piece on the back and wings.

† Norfolk name for the "Heron."

As the autumn advances, the old birds and their broods betake themselves to the salt-water firths, where they feed on any stranded fish that are cast up by the tide, or the wounded fowl that escape from the punt-gunners.

After gaining a living for a time in this manner, they no sooner observe a punt setting up to a flock of wild fowl, than, in expectation of a good dinner, they commence flying round the birds with loud screams, and not unfrequently, if hungry, darting down and spoiling the shot.

If plentiful, they generally become such constant attendants that it is impossible for the gunners to obtain a chance by day. On several occasions I have seen wounded fowl (both Mallard and Wigeon) fly from the attacks of the Gulls, and attempt to seek safety by returning and pitching in the water within a few yards of the punt. Small birds, such as Plover, are frequently carried off from under the very nose of the shooter, if falling in the water where they cannot readily be recovered.

The specimens in the case were obtained at a small sandy island, in Loch Shin, in Sutherland, in July, 1868.

GREAT BLACK-BACK GULL.—(Immature.)
Case 235.

The adult specimens, represented as attacking the Highland lamb, are only introduced to illustrate the destructive habits of this rapacious bird.

A weakly Ewe is no sooner discovered than she is set upon, and after being either forced into some crevice among the rocks, or slowly butchered by thrusts from their powerful bills, the Lamb next falls an easy victim.

This is by no means an uncommon occurrence, till this *interesting* Gull receives the benefit of the act for the preservation of sea birds.

The specimens in the immature stages, shewing the third or fourth, and also the second, years' plumages were shot in the Highlands, in **1877**.

MANX SHEARWATER.
Case 236.

I have frequently met with this bird during autumn in the North Sea, its curious flight always drawing attention, at whatever distance it might be seen.

The specimens in the case were shot in the Firth of Forth, August, 1874.

GREAT BLACK-BACKED GULL.—(Winter.)
Case 237.

During winter the snow white head and neck of the mature bird becomes slightly marked with grey, and the bright colours on the beak are less intense.

The specimen in the case was captured in a vermin trap, baited with a dead Pochard, that I had set for the benefit of the Grey Crows that came in flocks to consume the food of some Decoy Ducks I was keeping on Hickling Broad, in the winter of 1872.

COMMON GULL.—(Summer.)
Case 238.

Although it is stated as a fact in several ornithological works that this Gull nests in the cliffs on the sea-coast, I have never myself been so fortunate as to witness a single instance where this was the case; the whole of the breeding-stations that I have met with being either on islands, in fresh-water lochs, or on the open moors in the Highlands.

There is a colony on a small hill loch in Glenlyon, in the north-west of Perthshire, where I have frequently observed scores of these birds sitting on the tops of the trees, many of the highest branches being killed by their excrement. The nests are usually located among the rough stones and roots on the ground; but in two instances I have found them placed in the branches of the trees, though never at a greater height than about four feet.

This bird is most destructive to the Smolts, or juvenile Salmon, being often noticed feeding on the shallows of the rivers; it also preys on young birds, and is accused (though I have never myself observed it) of devouring the eggs of game.

The specimens in the case were obtained at the islands in the Lochs of Roro, in the north-west of Perthshire, in June, 1867.

COMMON GULL.—(WINTER.)

Case 239.

The mature bird in its winter dress, and the young in the immature stage, are shown in the present case.

Though feeding on fish, if it comes in their way, these birds during winter, when on the coast, appear to have a partiality for the mouths of sewers in the neighbourhood of large towns; here they are able to vary their diet with a choice selection of tit-bits: a crust of bread, half-an-inch of a tallow-dip, or a dead kitten, being each and all received with thanks.

They may also be noticed at this season following the plough, eagerly snatching up the worms and grubs as they are exposed to view.

The specimens in the case were shot at sea, off Brighton, during the winter of 1870.

OSPREY.
Case 240.

In several of the Northern counties of Scotland the Osprey may still be met with during the Summer months.

At the present time by far the greater number of nests are to be found on trees, while, if we are to believe old reports, the majority were placed on isolated crags of rock in fresh water lochs, and in a few instances on ruined castles.

The specimens were obtained in the Northern Highlands, in 1877.

KITE.
Case 241.

Though formerly of common occurrence all over the country, there are now but few counties in which the Kite can be reckoned as a resident.

I have noticed that the young birds of this species are usually supplied with a great variety of food; in nests I have examined there have been at different times several young rabbits and hares, a few squirrels and rats, numbers of grouse and peewits, and on two or three occasions the young of curlew, duck, and pigeons.

The specimens were obtained in the Northern Highlands, in 1877.

BLACK-HEADED GULL.—(Winter.)
Case 242.

In autumn, as soon as the young are sufficiently strong, these Gulls come down to the coast, where they appear to take up their residence for the winter, making occasional visits into the country in search of food

The Black-Headed Gull is partly nocturnal in its habits ; when stopping at Canty Bay, in East Lothian I have noticed small flocks coming regularly every evening, just at dusk, to certain spots along the shore, where they fed on the sand-hoppers and other insects which were plentiful among the dead sea-weed washed up by the tide.

The specimens in the case were shot on the coast of the Firth of Forth, in September and October, 1874.

BLACK-HEADED GULL—(Summer.)
Case 243.

This handsome bird is plentiful from north to south, breeding in colonies in various parts of the country.

Though frequenting the sea-coast during the autumn, winter, and early spring, it retires inland for the summer months, and rears its young, either in marshy districts or on the islands in some large piece of fresh water.

I was previously unaware that this species would destroy small birds; it was, however, recently stated (I suppose on good authority) in the " Zoologist," that such was occasionally the case. Its food, for the most part, I believe, consists of insects, beetles, worms, grubs, and moths, varied of course during the winter by marine animalcules and such refuse as it is able to pick up along the shore.

A large colony of these birds presents an exceedingly lively spectacle about the middle of June; owing to their nests having been, in many cases, frequently plundered, young of all sizes may be observed, from the small downy chick just breaking the egg-shell to the full-fledged bird that is taking its first flight.

As will be seen by the specimens in the case, the eggs of this species vary considerably.

The mature birds were obtained near Lairg, in Sutherland, in June, 1868, the young and eggs being taken at a small piece of water (known in Gaelic by a name that signifies the "Muddy Loch,") near Tain, in Ross-shire, in June, 1869.

GOOSANDER.

Case 244.

These birds make their appearance during severe weather, frequenting both fresh and salt water; the numbers, however, that arrive are very uncertain, depending greatly on the season, but few being observed should the winter prove mild.

If plentiful, they must be very undesirable visitors to a trout stream, as the quantity of fish they are capable of swallowing is something enormous.

The immature male and the female were shot in Perthshire, in 1867. The adult male in Inverness-shire, in March, 1878.*

LITTLE GULL.—(IMMATURE.)

Case 245.

This Gull is occasionally, during autumn and winter, far from uncommon on the eastern coast.

The specimen in the case was shot on Horsey Mere in the east of Norfolk, in November, 1871.

PUFFIN.

Case 246.

The home of this curious-looking bird is far from land, on the "rolling deep."

It is only for a few months during spring and summer that it approaches our shores for the purpose of rearing its young; unless wounded, or in some way disabled, it is seldom seen during winter.

The Puffin breeds in a variety of situations; holes in cliffs, rabbit-burrows, and ruined buildings being occasionally made use of.

The sketch from which the case is copied was taken from the old fortifications on the Bass Rock. The Puffin here nests in the holes in the crumbling masonry of the battlements.

The specimens were obtained in the Firth of Forth, in June, 1865.

---

* At the time the above notes were published I was not aware that the Goosander nested regularly in the British Islands.

Case **247.**

ROSEATE TERN.

Case **248.**

From various causes the numbers of these Terns that visit our shores are rapidly decreasing every season · though formerly nesting on several parts of the coast, but two or three stations are resorted to at the present time.

The specimen in the case was shot at the Fern Islands, in June, 1867.

MERGANSER.

Case **249.**

The breeding plumage of the male of this species is by no means so handsome as his winter dress; it cannot, however, be said that the cares of his family, or the toil of providing for their wants, have been the cause of his change, as the brightest of his colours were laid aside before the beginning of March.

These birds are particularly plentiful along the wild, rocky coasts of the northern parts of Scotland, breeding on the islands both off the coast and in the fresh-water lochs.

It is a singular habit with the present species, that two or three females, with their broods, frequently keep company; at times but one of the parents remaining with the school of young, while the others are following their own devices: this often gives rise to the idea that the family of the Merganser is more numerous than it really is.

The male and female were shot in Gairloch, off the west coast of Ross-shire, in May, 1868; the young being obtained on Loch Shin, in Sutherland, the following month.

GREAT CRESTED GREBE.—(Mature and Nestlings.)

Case 250.

After all the persecution that this curious bird has undergone for the sake of its feathers, it is a remark-able fact that it is still numerous in several parts of Great Britain, and though its haunts are rapidly becoming restricted by drainage, and other innovations, the Broads in the eastern counties, with their extensive reed-beds, are likely to afford it a safe asylum for several years to come.

During winter, even should the weather prove mild, the majority of these birds take their departure from their summer quarters, and are found for the most part singly, either on the tidal rivers or the open sea.

The summer plumage is acquired early in the year.

I have observed specimens with the frill or ruff fully developed in February and March, and on one occasion as soon as January 18th.

The nest of this species is a large accumulation of the stems of the reed amongst which it is built, the whole of the structure, together with its contents, being perpetually moist.

In three or four instances where I have closely watched the habits of this bird, I have discovered that the eggs were regularly laid on alternate days; the young I have also noticed being hatched at similar intervals.

The specimens in the case were obtained on Hickling Broad, in June 1871.

SANDWICH TERN.—(Mature—Summer.)
Case 251.

These fine birds are to be met with on several parts of the coast during the summer months. On the Fern Islands, off the coast of Northumberland, they nest in two or three different colonies in company with the Arctic Terns.

In a few localities in the north they breed on inland lochs.

The specimens in the case were obtained at the Fern Islands in June, 1867.

LESSER SPOTTED WOODPECKER.
Case 252.

The Lesser Spotted Woodpecker is, I believe, of much more frequent occurrence than is generally supposed, its small size enabling it to escape observation unless closely searched for.

I have particularly noticed that these birds, during the winter and early spring, appear to have a regular course that they pursue from day to day; having frequently timed them arriving in certain trees from the same direction almost to the minute, and after remaining their usual time among the branches, departing, if undisturbed, in the very line they had taken the previous day.

The male and female were obtained near Plumpton, in April, 1866, the young being taken from a pear-tree in a garden at Uckfield, in Sussex, in June 1872.

GREEN WOODPECKER.—(Mature.)
Case 253.

The present species is the commonest of our British Woodpeckers, being abundant in most wooded districts in the southern and midland counties of England. It, however, becomes scarcer towards the north and I have never met with it in Scotland.

The specimens in the case were shot in Balcombe Forest, in Sussex, in June 1875.

SANDWICH TERN.—(Mature and Immature.—Autumn.)
Case 254.

The old birds and their young are frequently noticed during autumn, in considerable parties, fishing along the coast of the Firth of Forth.

They also occasionally make their appearance at this season all round our shores.

By October they usually begin to make their way

south, but solitary examples are now and then met with during November.

The specimens in the case were shot in the Firth of Forth in August and September, 1874.

GREAT CRESTED GREBE.—(Mature and Half-grown Young.)
Case 255.

The present case shows the young Loons* between the age of four and five weeks.

It would hardly be supposed that a Perch of two or three inches long, with its prickly fin, would be a pleasant mouthful for a young bird of this age; the smaller specimen, however, contained a couple of fish of that length, while the larger had only a few feathers in its stomach.

The old and young were obtained on Hickling Broad in June 1873.

GREAT CRESTED GREBE.—(Immature.)
Case 256.

The specimens in this case are probably from ten to twelve weeks old.

While in the down the eye of this species is a light grey. During their first autumn and winter I have observed the iris both a dull orange and a bright lemon yellow; this gradually changes till it assumes the brilliant red which is seen in the adult in summer.

The birds were shot on Breydon, near Yarmouth, in September, 1871.

---

* **Norfolk name for the Great Crested Grebe.**

LESSER TERN.—(Mature.)

Case 257.

During the last fifteen or twenty years this graceful little bird has entirely disappeared from several of its breeding-stations in the south of England.

There are, however, still numerous localities where it may yet be found, though, I am afraid, in rapidly decreasing numbers.

The specimens in the case were shot at Rye, in Sussex, in May, 1860.

GREEN WOODPECKER.—(Mature & Immature.)

Case 258.

The hissing and snapping noise emitted by a brood of young Woodpeckers, when in the nest, would most probably deter anyone unacquainted with their note from exploring the hole in which they were concealed without due caution, as the sound would hardly be thought to proceed from a bird.

The specimens in the case were obtained in Balcombe Forest, in Sussex, in June, 1875.

At this early stage (the day the young first left the nest) it is already possible to distinguish the sexes, the black mark under the bill of the male containing a few red feathers, while that of the female is plain.

Case 259.

LESSER TERN.- -(Immature

Case 260.

Young birds in this stage are occasionally met with all round our coasts during the latter part of August and September.

The specimens were obtained near Hastings, in September, 1858.

MOORHEN.

Case 261.

This familiar bird, which is abundant from north to south, appears quite as contented on the smallest pool, or the narrowest drain, as on the most extensive lake in the United Kingdom.

Though not usually esteemed of much value as an article of food, both the Moorhen and its neighbour the Coot are, in my humble opinion, equal in flavour when properly dressed, to any bird that is found in the British Islands. Cooks, who are aware of the trouble of plucking them, are sure to declare they require to be flayed and then steeped in water, or possibly even perpetrate some such enormity as burying them in a cloth for several hours. Let them, however, simply be treated as Wild Duck or Teal, and I am almost certain that even the most fastidious epicure will be desirous of repeating the experiment. As the cook is the last person in an establishment that I should wish to be on bad terms with, I always, when shooting these birds, order the men to pluck those that are intended for home use as soon as they are collected ; while warm the feathers and down may be stripped from them with but little exertion, though, should they once get cold and set, it will require nearly an hour's scrubbing with resin to remove the tenacious black down from the breast and back of a single *Coot*.

The specimens in the case were obtained on Loch Slyn, near Tain, in Ross-shire, in May, 1869.

COOT.—(MATURE AND NESTLINGS.)
Case 262.

The *Coot* is a common bird in most parts of the British Islands, being found in numbers both on the Highland lochs and on the lakes and broads in the southern counties.

During winter they usually collect in flocks on some large piece of water ; here they remain till, driven from their quarters by continued frost, they come down to the tidal rivers and harbours on the coast, where numbers fall victims to the punt-gunners.

On the extensive broads in the eastern counties, where these birds begin to gather as early as September, regular battues are organised at intervals during the autumn and winter, and afford an exciting day's sport to the whole of the gunning community of the neighbourhood.

The meet is usually fixed for an early hour, but long before the appointed time the company may be seen arriving in craft of every description, those who are unable to get afloat contenting themselves by taking up a position on the banks. When all is ready the boats, numbering usually from thirty to fifty, form in line and work round the birds, so as to enclose them in a corner of the broad. As soon as they discover themselves hemmed in, they rise and fly in all directions, always at last making for the open water beyond the line of boats, affording great sport, and still greater confusion for several minutes.

If the line is well kept the *C*oots appear bewildered, and continue flying round and round for a considerable time before attempting to break through, but should only a single boat fall out of its appointed station, the whole of the birds in a body make for the gap, and the drive is spoiled.

Those that escape the first round settle on some remote corner of the broad, and are again attacked in the same manner.

The specimens in the case were obtained on Hickling Broad, in June, 1871. The young are a day or two old.

COOT.—(Mature and Half-grown Young.)

Case 263.

In this case the young are shown in a more advanced state, being probably six or seven weeks' old.

The specimens in the case were obtained in the Heigham Marshes, in the east of Norfolk, in June, 1873.

STORMY PETREL.

Case 264.

Though but seldom observed, except by seafaring people, the Stormy Petrel is common all round the British Islands.

I have often noticed these poor little birds terribly distressed by the buffetings they receive during a protracted gale, at times hovering and settling among the breakers, and occasionally being carried before some blinding squall, almost helpless, inland.

After a storm of several days' duration in November, 1872, I observed scores of these birds resting on the water a few miles off the coast of Norfolk, apparently thoroughly worn out, with their heads buried in their feathers. On visiting one of the lightships, I learned that several of the Stormy, as well as a single specimen of the Forked-tailed Petrel, had come on board while the gale was at its height.

It is late in the year before the Petrel arrives at its nesting-quarters, eggs, I believe, seldom being laid before the first or second week in June.

The birds in the case were shot in the English Channel, in May, 1872.

HAWFINCH.

Case 265.

The Hawfinch is frequently seen in considerable flocks in the south of England during winter. I have also noticed small parties at the same season in the midland counties, but have never myself met with this species in Scotland, though it is stated to have been occasionally observed.

Their roughly-built nest is commonly placed in fruit trees in gardens, and is usually so slightly constructed, that without close investigation it might readily be taken for a small collection of rubbish that had gathered among the branches.

Though feeding partly on beetles and insects, these birds are most destructive to peas; not content with eating what they require, they amuse themselves by slicing the pods with their powerful beaks, and of course destroying the contents.

The interior of the mouth of the nestling bird is of a most brilliant pale purple and cerise.

The specimens in the case were obtained near Plumpton, in Sussex, in June, 1872.

MARSH TIT.
Case 266.

The present is a widely distributed, though an exceedingly local species.

The specimens in the case were obtained near Plumpton, in Sussex, in March. 1866.

DUNLIN.—(Summer.)
Case 267.

The Dunlin breeds abundantly on the moors in the northern parts of the island. During autumn, winter, and early spring, they may be met with in smaller or larger flocks all round our shores, occasionally, after stormy weather, making their appearance inland.

About the middle of November, 1872, the coast of Norfolk was visited by a terrific gale, which, commencing on the Monday morning, raged with unabated fury during the whole week.

The first day the wind blew in furious squalls from north-north-east, and from daylight till about 3 p.m. a continued stream of Dunlins kept flying along the shore in face of the storm, flock after flock following one another in rapid succession for at least six hours.

I was able to distinguish a few Knots, Grey Plovers, Godwits, and Turnstones among them, but the numbers of these birds were trifling compared with the Dunlins. The following day several scattered flocks were still passing north, and till the end of the week a few were noticed flying in the same direction.

Those naturalists who complain that our smaller waders are on the decrease, would, I think, have had their minds set at rest had they been for half-an-hour only within sight of Yarmouth beach, on Monday, the 11th of November, 1872.

The specimens in the case were obtained on " the Fendom," a flat, sandy waste in the neigbourhood of Tain, in Ross-shire, in June, 1869.

DUNLIN.—(WINTER.)

Case 268.

During winter these birds occasionally collect in immense flocks, at times keeping company with Knots and other waders.

The specimens in the case were shot between Shoreham and Lancing, in Sussex, in January, 1870.

PIGMY.—(Mature—Autumn and Winter.)

Case 269.

The Curlew Sandpiper, or Pigmy, as it is more commonly called, is here represented in both autumn and winter plumage.

During autumn mature birds are not uncommon on the eastern coast, being usually found in small parties of from two or three to half-a-dozen, occasionally, though very seldom, joining with the immature birds of their own species, appearing rather to prefer the company of Dunlins or Knots.

I have never met with this bird between November and the latter end of April; the specimen that is shown in its full winter dress being singularly enough killed on Breydon mudflats, on the 26th of May, 1871, the remainder were shot the following autumn in the same locality.

DUNLIN.—(Immature.)

Case 270.

The immature birds are here shown in their first feathers. In this stage they frequently make their appearance on the mudbanks in the south as early as the end of July or beginning of August.

The specimens in the case were shot in Shoreham Harbour, in August, 1872.

COLE TIT.

Case 271. -

The Cole Tit is common from north to south, being particularly abundant among the fir plantations in the Highlands.

The specimens were obtained in the Tarlogie Woods, near Tain, in Ross-shire, in June, 1869.

CHAFFINCH.

Case 272.

Though the Chaffinch is a resident at all times and seasons in the British Islands, numbers arrive in the autumn from the North of Europe, and having had wings sent me from the lightships during the early spring, I conclude our visitors depart again at that season.

The specimens in the case were obtained in the neighbourhood of Brighton, in May, 1872.

GOLDFINCH.

Case 273.

Thanks to the birdcatchers, Goldfinches have greatly decreased in the last twenty years. It is seldom that a flock of from fifteen to twenty is met with at the present time, where formerly the birds could be seen in hundreds.

The specimens in the case were obtained at Hickling, in Norfolk, in June, 1873.

CRESTED TIT.
Case 274.

Seldom straying far from its native haunts on the wild fir-covered hills that are situated in the central and eastern parts of the Highlands, the Crested Tit is generally supposed to be a scarcer bird than it really is.

While watching the brood that are shown in the case, I observed that the female procured the whole of the insect food with which she supplied the young from among the stems of the juniper bushes. The male appeared to interest himself but little in the domestic duties.

The specimens were obtained near Inverness, in June, 1876.

PIGMY.—(Summer.)
Case 275.

The Pigmy in summer plumage is but seldom obtained in the British Islands.

At this season the bird appears somewhat shy and wary; I have, when shooting on the Norfolk mud-banks during May, frequently observed one or two at a distance, though I never was lucky enough to procure but the pair in the case.

The specimens were killed on Breydon, in May, 1871.

PIGMY.—(Immature.)
Case 276.

Large flocks of immature birds are occasionally met with in suitable localities, during autumn, all round our coasts.

The Nook at Rye Harbour, on the Sussex coast, was formerly one of their most favourite feeding-grounds; grass marshes, however, were preferred by the owners of the land to sea-water mudbanks, and a wall having been built, the tide was at last successfully kept back. I happened, the first autumn after the alteration, to be present, when a large mixed flock of Pigmies and Stints, after wheeling round two or three times, settled down among the sheep, which were now the occupants of their former quarters; after running about in the grass for a time, apparently bewildered by the alteration that had taken place in the nature of the soil, they at last became acquainted with the state of affairs, and uttering a shrill cry, the whole flock took wing, evidently disgusted with the transformation that had been effected.

The specimens were obtained at Rye, in Sussex, in September, 1858.

WHIMBREL.

Case 277.

A few pairs of these birds still breed in the wilder parts of the north of Scotland, and on some of the adjacent islands; several of the localities, however, that they formerly resorted to are entirely deserted.

During spring and autumn considerable flocks may be met with all round our coasts.

The specimens in the case were shot at Rye, in Sussex, in May, 1862.

PEEWIT.

Case 278.

This species is abundant in the British Islands from north to south, though its numbers would doubtless be far greater, were it not for the persecution it undergoes by being robbed of its eggs, which, unfortunately for the bird, have a great reputation as a dainty dish.

In autumn and winter Peewits collect in large flocks, and make their appearance on the mudbanks in the neighbourhood of the coast.

I do not know whether it is generally allowed that the numbers of our native birds are increased during the winter by arrivals from abroad; I have, however, on two separate occasions observed large flocks in the North Sea, at a considerable distance from land, evidently making for some part of our eastern coast.

The specimens in the case were obtained on Gullane Links, in East Lothian, in May, 1867.

GARDEN WARBLER.

Case 279.

The Garden Warbler, like the Black Cap and Nightingale, is only a visitor to our shores, being found during summer in most of the English counties; its unobtrusive habits and sober colouring lead, however, to its being frequently overlooked.

The specimens in the case were obtained in the neighbourhood of Brighton early in the autumn of 1869.

SISKIN.

Case 280.

A few of these birds may be met with nesting in some of the northern counties of Scotland ; as a rule, however, the flocks that visit us during the winter take their departure on the approach of spring.

The specimens in the case, together with their nest and eggs, were obtained in the Tarlogie Woods, near Tain, in Ross-shire in June, 1869.

GREENFINCH.

Case 281.

The present species is one of our commonest birds, being met with in almost every part of the British Islands where the land is cultivated.

The specimens in the case were obtained at Portslade, near Brighton, in June, 1872.

TEMMINCK STINT.—(SUMMER.)

Case 282.

It is by no means a common occurrence for this species to be met with in Great Britain during the Summer months. Though occasionally observed in the Spring, while on the way to its breeding grounds in Lapland, it has at that time hardly acquired the full Summer plumage.

The specimen was shot on the mud-flats in Shoreham Harbour, in July, 1878.

WATER RAIL.

Case 283.

The Water Rail is usually met with in low-lying marshy ground, occasionally, however, being found frequenting the banks of rough streams in wooded districts.

The extensive reed-beds and trackless swamps that abound in the neighbourhood of the Broads in the eastern counties are admirably adapted to their habits, and the bird, though but seldom seen, may be heard continuously during fine still weather in the summer months; the peculiar squeaks and grunts that are emitted by this species would never be supposed by anyone unacquainted with its note to proceed from the throat of a bird.

The specimens in the case were obtained in the Potter Heigham marshes, in Norfolk, in May, 1870.

SPOTTED RAIL.

Case 284.

Though the Spotted Rail is usually considered a summer visitor to our shores, a few, I believe, will occasionally remain during the winter in suitable localities.

Like its neighbour the Water Rail, the present species is particularly abundant in the Broad district in the eastern counties. While snipe shooting in the autumn, I have repeatedly found them in great numbers in the neighbourhood of Yarmouth.

The male and female were obtained in Pevensey Marsh, in April, 1866, the nest and eggs being taken in a reed-bed near Hickling Broad, in June, 1873.

SCLAVONIAN GREBE.—(Summer.)

Case 285.

I have met with this Grebe at various times during spring and winter all round our shores, from Sussex to Sutherland.

The specimens in the case were shot in Ross-shire, in April, 1869 ; the male being killed on Loch Slyn and the female on the Dornoch Firth. A perfectly snow-white Grebe, which I believe to have been of this species, was in company with the male when that bird was obtained, but the day being exceedingly stormy, it was lost sight of in the broken water.

LAND RAIL.

Case 286.

The Land Rail, or Corncrake, is widely distributed over the British Islands, being remarkably plentiful in the cultivated portions of some of the Highland glens, though, as it generally takes its departure before the crops are sufficiently cleared to permit of shooting, its presence, owing to its skulking habits, would seldom attract attention were it not for its monotonous croaking note. I have at times fallen in with these birds in great numbers while Partridge-shooting in Sussex early in September, on one occasion bagging eight and a half couple in about two acres of clover.

I was greatly surprised one summer, when fishing on the Lyon, in Perthshire, to see a Land Rail, which my retriever had disturbed, run down to the bank of the river, and, without pausing a moment, drop quietly

into the water and strike boldly out for the opposite
shore ; in less than a minute the dog arrived on the
bank, and catching sight of it immediately captured it
in the water before it had time to gain the land.

The bird, which I examined alive, had not received
the slightest injury, being blessed with the full use of
both wings and legs, so that its taking to the water was
entirely a matter of choice.

The specimens in the case were obtained in Glenlyon,
in Perthshire, during the summer of 1867.   The young,
which were captured by the retriever, must, I should
imagine, have been a second brood, being taken as late
as the 1st of September.

RED-BACKED SHRIKE.

Case 287.

The Butcher-bird, as this species is more frequently
styled in the south, arrives in the beginning of May,
and, after rearing its young, departs early in the
autumn.

In some parts of Sussex, and also in the grass
country in the neighbourhood of Harrow-on-the-Hill,
a few miles north of London, this bird is particularly
abundant.

I have never myself observed them further north
than Norfolk, though they occur in Yorkshire, and
have at times, it is said, been met with in Scotland.

They prey on beetles and other large insects, at
times, for convenience in feeding, transfixing them on
thorns in hedges. I once noticed a male flying with
what appeared to be an old Yellowhammer in his
claws: though it is stated that the Shrike occasionally

destroys young birds, I should hardly have imagined one capable of slaying a full-grown Yellow Bunting.

The female and young were obtained at Potter Heigham, in Norfolk, in July, 1869; the male being killed in Sussex, on his first arrival in May, at which season the plumage is always in its greatest perfection.

REDPOLE.

Case 288.

This lively little bird breeds plentifully in the Highlands of Scotland, and also in several of the northern and midland counties of England.

I have seen nests in the neighbourhood of Brighton, but from the appearance of the parent birds I should judge that, in every instance, they had escaped from confinement.

During winter they are found in large flocks in all parts of the country.

For some time I watched the specimens that are in the case engaged in building their nest, which was placed in an alder bush close to a stream, and lined with the white floss that forms the flower of the cotton grass. I observed that the female performed the whole of the work, collecting the materials, and also working them together, the male at times accompanying her while gathering them, but simply amusing himself by flitting from twig to twig without offering the slightest assistance.

The male and female were obtained at Potter Heigham, in Norfolk, in June, 1873, the young having been caught in some gardens in Brighton, in August, 1869.

MEALY REDPOLE.

Case 289.

This species is only an autumn and winter visitor to
our shores, at times appearing in considerable flocks.

I have known it plentiful in Sussex, frequenting
the alder bushes in the interior of the county, and
also being occasionally met with in the immediate
neighbourhood of Brighton. In Norfolk, the vicinity
of Norwich appears a most favourite locality for this
bird. It is also at times observed in various parts of
the British Islands.

The specimens in the case were obtained among
the alder trees on the banks of the Heigham river,
near Norwich, in December, 1873.

OSPREY.—(IMMATURE.)

Case 290.

It is seldom that a season passes without a
specimen or two of this species being either seen
or obtained in the Southern or Eastern Counties ;
the rivers and ponds of Sussex and the broads
of Norfolk appear to be particularly favourite
resorts.

The Osprey is probably three or four years old
before it pairs and nests; those that are observed
in the South during the Summer are still in the
immature state.

The bird, which is in the second or third years'
plumage, was shot with a punt gun while perched
on a stake on Breydon Mud Flats, in May, 1871.

TEMMINCK'S STINT.—(IMMATURE.)

Case 291.

A few of these birds may generally be met with every autumn, in suitable localities, all round our southern and eastern coasts.

I have observed them at times in the neighbourhood of the fresh-water broads in Norfolk, and have also found them frequenting the mudbanks of Breydon, and the flat harbours on the Sussex coast.

The specimens in the case were obtained on Breydon mudflats, in September, 1872.

SCLAVONIAN GREBE.—(WINTER.)

Case 292.

The visits of this Grebe are usually more numerous during the winter than at any other season.

The specimens in the case were shot on Heigham Sounds, in Norfolk, in December, 1871.

LITTLE GREBE.—(SUMMER.)

Case 293.

This small diver is plentiful in Great Britain, being found during summer from north to south. I have seen as many as three or four pairs engaged with their broods on some of the smaller lochs in Ross-shire, and have also met with them in both Sutherland and Caithness. In England, it may be said to occur in almost every county; it is, however; strange that in the larger broads in Norfolk, where there are endless reed-beds and swamps, apparently adapted to its habits, the bird is by no means common.

The specimens in the case were obtained at a small muddy loch near Tain, in Ross-shire, in June, 1869.

TEMMINCK'S STINT.—(Winter.)

Case 294.

The mature bird, in its winter dress, is but seldom observed in this country.

The specimens in the case were shot on Breydon mudflats, in September, 1872.

WHINCHAT.

Case 295.

During the summer months the Whinchat may be found widely distributed over the British Islands.

On his first arrival in the spring, the male is a bright, handsome bird ; at the time of his departure, however, in the autumn, he can hardly be distinguished from his plainly-dressed family.

The specimens in the case (with the exception of the male, who was shot earlier in the season) were obtained at Potter Heigham, in Norfolk, in July, 1870.

HOUSE SPARROW.

Case 296.

The present case is copied from a sketch made at Falmer, near Brighton, where a Sparrow's nest was placed in a hole among the crumbling chalk and mould in an overhanging bank.

The specimens were obtained in the neighbourhood of Brighton, in June, 1872.

TREE SPARROW.

Case 297.

This bird is common in the east of Norfolk, in a few localities being equally as numerous as its relative the House Sparrow.

In Sussex, I have occasionally seen large flights pass over during winter; and in April, 1875, I noticed several with a large flock of Bramblings and Chaffinches that remained for several weeks feeding in the fields near Falmer. I have not, however, observed the Tree Sparrow breeding in this county.

Large flocks arrive from the north of Europe in the autumn. On several occasions I have met with them in tho North Sea many miles from land, at times appearing much fatigued, and remaining on board for several hours to rest.

The following statement, in reference to the present species, appears in a well-known ornithological work:—

"It is now perfectly clear that this bird resides amongst trees only, and that it makes its nest in holes and cavities of such as are decayed, and never amongst the branches nor in buildings."

I took particular trouble to hunt for the breeding quarters of these birds round several farms in the east of Norfolk, and in every instance the nest was placed amongst the buildings; some in cowsheds, others under the tiles of the out-houses, and three or four

among the rough stems of some particularly coarse
ivy that grew over an old wall.   Not one did I discover
amongst the trees, though the House Sparrows were
breeding plentifully both in the branches and the ivy
round the trunks.

The male and female in this sparrow are alike ; the
young also exhibit the same markings in their first
feathers.

The specimens in the case were obtained at Potter
Heigham, in Norfolk, in June, 1873.

STONE *CHAT*.—(SUMMER.)
Case 298.

The Stone Chat is widely distributed over the British
Islands, frequenting open heaths and furze-covered
downs.

A few remain with us during the winter, but their
numbers are increased in the spring by arrivals from
the Continent.

The specimens in the case were obtained on the
Downs, near Brighton, in June, 1872.

LITTLE STINT.—(SUMMER.)
Case 299.

This elegant little wader (which both in habits and
appearance takes after the Dunlin, while the
Temminck's Stint more closely resembles the common
Sandpiper) is found on several parts of our coasts
during spring and autumn.

The numbers, however, that are met with are far

greater at the latter season, when the majority of the flocks are composed of immature birds.

Two of the specimens in the case were shot on Hickling Broad, in June, 1870, and the remainder on Breydon mudflats, in May, 1871.

LITTLE GREBE.—(Winter.)
Case 300.

This case shows the Little Grebe (or Dabchick, as it is frequently called) in its winter dress.

The specimens were obtained in the neighbourhood of Shoreham, in Sussex, in November, 1870.

RED-NECKED GREBE.
Case 301.

The Red-Necked Grebe is of uncertain occurrence in the British Islands; the majority of the specimens procured are, however, obtained along the eastern coast.

The bird in the case exhibits the immature plumage, and was shot on Breydon mudflats, in August, 1873.

LITTLE STINT.—(Autumn.)
Case 302.

From August to October these birds are occasionally found in suitable localities all round our coasts.

Before the mudbanks at Rye Harbour, in Sussex, were drained, large flocks of Stints used annually to visit the spot; the change that has been effected has, however, considerably reduced their numbers, though small parties, I believe, still make their appearance in the neighbourhood of their former haunts.

The specimens in the case were shot in the back-water at Rye Nook, in Sussex, in September, 1858.

STONE CHAT.—(Autumn.)

Case 303.

The present case shows the more sober colouring that this species exhibits during the autumn and winter months.

The specimens were shot between Shoreham and Lancing, in Sussex, in September, 1875.

BULLFINCH.

Case 304.

This well-known bird is met with in most parts of the British Islands.

Though apparently rather out of their latitude, I observed a small flock one winter amongst the stunted fir-trees and rough stones near the summit of the hills to the north of Glenlyon, in Perthshire; the Bullfinch may, however, be found during summer in most of the Highland glens where there is sufficient cover to provide them with nesting quarters.

The specimens in the case were obtained near Plumpton, in Sussex, in January, 1870.

GOLDEN EAGLE.

Case 305.

The Golden Eagle is stated by several writers t
be rapidly disappearing from the British Islands
there is, however, I am well convinced, but littl
fear that the bird is in any danger of bein
exterminated for many years to come.

If the Eagles of former days showed an
bravery in the defence of their nests and young
I am afraid the race has sadly degenerated, a
more cowardly brutes than those specimens o
the "*noble bird*" that have come under my ow
observation would be hard to find.

The case is copied from a nest in Sutherland.
shire.

The specimens were obtained in the Wester
Highlands in the Spring of 1877.

GOLDEN EAGLE.—(IMMATURE.)

Case 306.

The Eagles are here shown in different stages o
plumage. The bird on the left hand is almos
adult; the other, in the immature state, probabl
just a year old.

Both specimens were trapped in the Wester
Highlands in the Spring of 1877.

KITE.—(NESTLING.)
Case 307.

The nest of the Kite is a curious collection of rubbish. Remnants of old flannel garments, scraps of paper, sheep's wool, and horse hair, being used to line a cradle composed of dead twigs of fir and other trees.

The young bird (just hatched) and the eggs, together with the nest, were obtained in Ross-shire, in 1877.

NORTHERN WILLOW WREN.
Case 308.

This bird, which closely resembles " Gaetkes Warbler," is said, by scientific Naturalists who have examined it, to be the form of Willow Wren that is usually found in the Arctic circle.

I shot the bird, my attention being drawn by its curious note, near Brighton, in April, 1876.

RING GUILLEMOT.—(WINTER.)
Case 309.

The Ring Guillemot is not unfrequently met with in the English Channel during Winter and early Spring.

The specimen was shot off Brighton in 1877.

GOOSANDER.—(YOUNG.)
Case 310.

During the Summer of 1878 I met with several broods of Goosanders in various parts of the Northern Highlands.

The specimens were shot near Inverness in July, 1878.

BLACK BIRD.—(Immature.)
Case 311.

The specimens, which were obtained near Brighton in 1877, shew the light colour of the plumage of the young birds before their first moult.

CANADA GOOSE.
Case 312.

Some writers consider it doubtful if any of the birds of this species obtained in Great Britain are really wild, believing them to have escaped from confinement.

The specimens were shot on the Norfolk coast early in 1878.

RING DOTTEREL.—(Immature.)
Case 313.

Great variety in size and colour is to be observed among the birds composing the large flocks of Dotterel frequenting the South-Coast during the Autumn.

The specimens were obtained at Shoreham in September, 1878.

GREEN SANDPIPER.—(Summer.)
Case 314.

It is but seldom that this species is met with in Great Britain during the Summer.

The specimen was shot on the marshes near Hickling Broad, in Norfolk, in June, 1873.

POLISH SWAN.
Case 315.

This bird is considered by scientific writers to be a distinct species from the Mute Swan.

The specimen was shot in Norfolk, in the Autumn of 1873.

CPSIA information can be obtained at www.ICGtesting.com
Printed in the USA
LVOW10s0945091215

465966LV00028B/1246/P